Analysis of Integrated Data

Chapman & Hall/CRC
Statistics in the Social and Behavioral Sciences Series

Series Editors

Jeff Gill
Washington University, USA

Steven Heeringa
University of Michigan, USA

Wim J. van der Linden
Pacific Metrics, USA

J. Scott Long
Indiana University, USA

Tom Snijders
Oxford University, UK
University of Groningen, NL

Aims and scope

Large and complex datasets are becoming prevalent in the social and behavioral sciences and statistical methods are crucial for the analysis and interpretation of such data. This series aims to capture new developments in statistical methodology with particular relevance to applications in the social and behavioral sciences. It seeks to promote appropriate use of statistical, econometric and psychometric methods in these applied sciences by publishing a broad range of reference works, textbooks and handbooks.

The scope of the series is wide, including applications of statistical methodology in sociology, psychology, economics, education, marketing research, political science, criminology, public policy, demography, survey methodology and official statistics. The titles included in the series are designed to appeal to applied statisticians, as well as students, researchers and practitioners from the above disciplines. The inclusion of real examples and case studies is therefore essential.

Recently Published Titles

Generalized Linear Models for Categorical and Continuous Limited Dependent Variables
Michael Smithson and Edgar C. Merkle

Incomplete Categorical Data Design: Non-Randomized Response Techniques for Sensitive Questions in Surveys
Guo-Liang Tian and Man-Lai Tang

Handbook of Item Response Theory, Volume One: Models *Wim J. van der Linden*

Handbook of Item Response Theory, Volume Two: Statistical Tools *Wim J. van der Linden*

Handbook of Item Response Theory, Volume Three: Applications
Wim J. van der Linden

Computerized Multistage Testing: Theory and Applications
Duanli Yan, Alina A. von Davier, and Charles Lewis

Applied Multivariate Analysis in the Behavioral Sciences, Second Edition
Kimmo Vehkalahti and Brian S. Everitt

Analysis of Integrated Data
edited by Li-Chun Zhang and Raymond L. Chambers

For more information about this series, please visit: https://www.crcpress.com/go/ssbs

Analysis of Integrated Data

Edited by
Li-Chun Zhang
Raymond L. Chambers

CRC Press
Taylor & Francis Group
Boca Raton London New York

CRC Press is an imprint of the
Taylor & Francis Group, an **informa** business
A CHAPMAN & HALL BOOK

CRC Press
Taylor & Francis Group
6000 Broken Sound Parkway NW, Suite 300
Boca Raton, FL 33487-2742

First issued in paperback 2021

© 2019 by Taylor & Francis Group, LLC
CRC Press is an imprint of Taylor & Francis Group, an Informa business

No claim to original U.S. Government works

ISBN 13: 978-1-03-209278-2 (pbk)
ISBN 13: 978-1-4987-2798-3 (hbk)

This book contains information obtained from authentic and highly regarded sources. Reasonable efforts have been made to publish reliable data and information, but the author and publisher cannot assume responsibility for the validity of all materials or the consequences of their use. The authors and publishers have attempted to trace the copyright holders of all material reproduced in this publication and apologize to copyright holders if permission to publish in this form has not been obtained. If any copyright material has not been acknowledged please write and let us know so we may rectify in any future reprint.

Except as permitted under U.S. Copyright Law, no part of this book may be reprinted, reproduced, transmitted, or utilized in any form by any electronic, mechanical, or other means, now known or hereafter invented, including photocopying, microfilming, and recording, or in any information storage or retrieval system, without written permission from the publishers.

For permission to photocopy or use material electronically from this work, please access www.copyright.com (http://www.copyright.com/) or contact the Copyright Clearance Center, Inc. (CCC), 222 Rosewood Drive, Danvers, MA 01923, 978-750-8400. CCC is a not-for-profit organization that provides licenses and registration for a variety of users. For organizations that have been granted a photocopy license by the CCC, a separate system of payment has been arranged.

Trademark Notice: Product or corporate names may be trademarks or registered trademarks, and are used only for identification and explanation without intent to infringe.

Publisher's Note
The publisher has gone to great lengths to ensure the quality of this reprint but points out that some imperfections in the original copies may be apparent.

Library of Congress Cataloging-in-Publication Data

Names: Zhang, Li-Chun, editor. | Chambers, Raymond L., 1950- editor.
Title: Analysis of integrated data / edited by Li-Chun Zhang, Raymond L. Chambers.
Description: Boca Raton, Florida : CRC Press, [2019] | Includes bibliographical references and index.
Identifiers: LCCN 2019006742| ISBN 9781498727983 (hardback : alk. paper) | ISBN 9781315120416 (e-book : alk. paper)
Subjects: LCSH: Multivariate analysis. | Multiple imputation (Statistics) | Measurement uncertainty (Statistics) | Mathematical statistics.
Classification: LCC QA278.7 .A53 2019 | DDC 519.5/35--dc23
LC record available at https://lccn.loc.gov/2019006742

Visit the Taylor & Francis Web site at
http://www.taylorandfrancis.com

and the CRC Press Web site at
http://www.crcpress.com

To Till and Elise and to Pat

Contents

Preface

The rapid uptake of administrative and transactions data use in official, scientific and commercial applications in the last decade, as well as the impact of the "Big Data Revolution", is leading to a growing diversification of primary data sources. This has resulted in an increasing interest in methods for statistical analysis that account for the fact that such data are often collected for reasons that have nothing to do with the application of interest.

Methods for assessing the representativeness of data drawn from a single source – such as with responses obtained from a sample survey, or treatment results collected in an observational study, or sentiment measures recorded by a group of social media users – are reasonably well understood, as are methods for dealing with the impact of any additional errors potentially caused by the use of an imperfect measurement instrument (or mechanism). However, when data from multiple sources are combined to enable statistical inference, or to generate new statistical data for purposes that cannot be served by each source on its own, an additional source of error emerges, corresponding to whether the data obtained from the study units in the combined data are, in fact, valid observations from the population of measurements that these data are supposed to represent.

For an overarching perspective, one may refer to this problem as the problem of inference under *entity ambiguity*. From a data integration perspective, this problem arises whenever it is not possible to state with certainty that the integrated source corresponds to the target population of interest. In other words, a situation of entity ambiguity can be characterised by the *lack* of an identified population set of target units or an observed subpopulation set of such units. This book covers a range of topics encountered in three generic settings of entity ambiguity: statistical analysis and uses of linked datasets that may contain linkage errors; datasets created by a data fusion process, where joint statistical information is simulated using the information in marginal data from non-overlapping sources; and estimation of target population size when target units are either partially or erroneously covered in each source.

In this book, we emphasise the statistical uncertainty and inference issues that arise when there is entity ambiguity in a dataset created by integrating information from multiple sources. In doing so, our aim is to provide readers with a solid theoretical basis for the analysis of integrated data, with a particular focus on applications in the social sciences and in the public and private sectors. Hopefully, the issues raised, as well as the problems tackled, in the chapters that make up this book will provide inspiration for future research in this important topic area.

The genesis for this book goes back to 2008, when Ray was developing his ideas on regression analysis of probability-linked data, and Li-Chun visited him in Wollongong. This inspired Li-Chun to develop a unit-error theory for households created in a statistical register. Much later, when Ray visited Li-Chun in Southampton in the summer of 2014, both of us were wrestling with issues of statistical inference under data fusion. Our discussions at that time gradually evolved into the theory of minimal inference, and also led us to propose a book project to Chapman & Hall/CRC Press focussing on the analysis of integrated data.

It is a pleasure to acknowledge the support we have received in bringing this project to completion. First, we are most grateful to all the contributors for sharing their work with us. It goes without saying that the book would not have been possible without their contributions. However, we must also confess that, as the editors of this book, it is even more pleasing that the contributors are friends and colleagues, with whom we have cooperated in various international projects and exchanged ideas on numerous occasions. Li-Chun's work with this book has also benefitted from grants by the UK Economic and Social Research Council's Administrative Data Research Centre England, while Ray's involvement has benefitted from research commissioned by the Australian Bureau of Statistics, and also from his participation in the Economic and Social Research Council's Longitudinal Studies Review 2017. We would like to acknowledge the Research for Official Statistics in Europe (ROSE) partnership, with inaugural funding from the University of Southampton's Global Partnerships Award, which provided support for many of the authors who contributed to this book. Finally, we thank the publishers, Chapman & Hall/CRC Press and in particular Rob Calver and Lara Spieker, for their support and patience over the long time that it took us to put together the manuscript for this book.

Li-Chun Zhang and Ray Chambers
Southampton and Wollongong, March 2019

Contributors

Bart F. M. Bakker
Statistics Netherlands
Den Haag
and
VU University
The Netherlands

Roberto Benedetti
Department of Economic Studies
University "G. D'Annunzio" of
    Chieti-Pescara
Pescara, Italy

Raymond L. Chambers
University of Wollongong
Wollongong, Australia

Pier Luigi Conti
Sapienza University of Rome
Rome, Italy

Maarten Cruyff
Utrecht University
Utrecht, The Netherlands

Davide Di Cecco
Istat, National Institute of Statistics
Rome, Italy

Loredana Di Consiglio
Istat, National Institute of Statistics
Rome, Italy

Marcello D'Orazio
Istat, National Institute of Statistics
Rome, Italy

Marco Di Zio
Istat, National Institute of Statistics
Rome, Italy

Maria Michela Dickson
Department of Economics and
    Management
University of Trento
Trento, Italy

Giuseppe Espa
Department of Economics and
    Management
University of Trento
Trento, Italy

Danila Filipponi
Istat, National Institute of Statistics
Rome, Italy

Diego Giuliani
Department of Economics and
    Management
University of Trento
Trento, Italy

Daniela Marella
Università Roma TRE
Rome, Italy

Federica Piersimoni
Istat, National Institute of Statistics
Rome, Italy

Mauro Scanu
Istat, National Institute of Statistics
Rome, Italy

Paul A. Smith
University of Southampton
Southampton, UK

Tiziana Tuoto
Istat, National Institute of Statistics
Rome, Italy

Peter G. M. van der Heijden
University of Southampton
Southampton, UK

and

Utrecht University
Utrecht, The Netherlands

Joe Whittaker
University of Lancaster
Lancaster, UK

Li-Chun Zhang
University of Southampton
Southampton, UK

and

Statistics Norway University of Oslo
Oslo, Norway

1

Introduction

Raymond L. Chambers

University of Wollongong, Wollongong, Australia

CONTENTS

1.1 Why this book?

Over the last twenty-five years, there has been an explosion in computing power that has made the creation and interrogation of large databases covering large national populations comparatively straightforward. Previously, access to such data sources with comprehensive population information, often referred to as population registers, had realistically only been available for Northern European populations with a long tradition of public acceptance of government access to such data via unique personal identifiers. But now population level data are collected by a huge number of organisations, both government and non-government, for a wide variety of purposes, and these data are often in the public domain, or at least available on request to people outside the organisations responsible for collating them. This has created many new opportunities for researchers in official statistics and in academia to access potentially extremely useful data for investigating population relationships. These opportunities are often referred to as part of the "Big Data Revolution".

In the second half of the twentieth century, the main method of research data acquisition for research and policy making was the sample survey, and the data collected in well-designed surveys underpinned many advances in social, economic and environmental science. However, survey statisticians, and particularly those that worked on official censuses and surveys, recognised from an early stage that the opportunities presented by rapidly increasing computing power, both for data processing as well as data storage, implied that the role of such traditional data gathering exercises would change rapidly. Key drivers for this change have been the rapidly increasing costs (included perceived response burden) of carrying out national surveys, and their increasing rates of non-response and non-participation. As a consequence, there is now a greater willingness to regard sample surveys as one among many ways of collecting and analysing data to inform science and social policy. In this new, and rapidly developing world of complementary data sources, survey and census methodologists are now also changing to accommodate the new design and inference problems that inevitably arise. This is particularly the case when traditional surveys and censuses are integrated with external data sources, or even replaced by them. The common thread here being the need to develop design and inference methods for a future where official statistics may well be "assembled to order" from a variety of data sources, with surveys not necessarily constituting one of these sources.

It is well known that common sampling design features, such as stratification, multistage and unequal probability selection, generally complicate the distribution of sample survey data. Standard statistical methods of analysis, such as regression and hypothesis testing, therefore need to be adapted, or redefined, when based on such data. But there is also a growing need to reduce both the cost and burden associated with the purposeful, targeted data collection implicit in a scientific survey. The same considerations about the impact of data provenance on data analysis apply when statistical analysis is based instead on data derived from a variety of sources, and particularly where these sources correspond to databases that are generated, recorded and/or maintained for reasons that are often unrelated to the analysis of interest. That is, the flip side of the rapidly increasing demand for statistical evidence that cannot be met by the data directly collected using designed sample surveys and censuses is the need to develop methods of analysis that can accommodate the complex data structures that arise when these data are sourced instead from registers that have typically been created with quite different information needs in mind.

Data integration involves combining data residing in different sources to enable statistical inference or to generate new statistical data for purposes that cannot be served by the data in their initial state or from each source on its own. The approach can yield significant gains for scientific as well as commercial investigations, and is necessary when sharing data from multiple sources, including surveys. As a consequence, there is now a clear trend within official statistics for data to be collected via integration, in contrast to the intensive and costly direct data collection for statistical purposes that has traditionally been the undisputed official statistics standard.

In response to this changing set of data collection priorities, a framework for "analysis of integrated data" is required as a natural extension to "analysis of survey data". In addition to possible sampling error and the measurement errors associated with the data sources, integrated data can have many distinct types of errors of its own, depending on the integration process. It is therefore generally inappropriate to treat the integrated data as is, and to apply standard methods of analysis without modification. There is no free lunch when data sources are combined for analysis – inferential methods that account for the way in which the combining has been carried out are necessary.

1.2   The structure of this book

This book addresses methods of statistical analysis involving distinct, sometimes complex distributional assumptions that are necessary for integrated data obtained from multiple sources. The central theme is a two-fold one: (1) how to adapt statistical methods of analysis in the presence of the various potential errors inherent to integrated data; (2) how to develop appropriate methods of inference that are made possible given the integrated data. The book itself is organised into three parts, each made up of chapters that address a specific set of analysis issues that have been observed in an integrated data environment.

Part 1: Uncertainty due to errors in record linkage and statistical matching

Data linkage is often referred to as record matching or record linkage, and has long been a strategy for gathering information about the same individual by combining data from two or more distinct sources. Dunn (1946) defined it as a process used to gather facts for a "book of life", i.e., a record of events occurring and recordable, experienced by the same individual between birth and death. Since then there have been many definitions for data linkage, e.g., "Data linkage is the joining of two or more administrative or survey datasets using individual reference numbers/identifiers or statistical methods such as probabilistic matching" (ADRN, 2012). Computerised record linkage is ubiquitous in most areas of commerce, administration and science, with automatic matching of very large datasets now quite straightforward. This is particularly the case when there are common variables (identifiers) stored on these datasets, which in many cases is a consequence of the move to digitised service provision and more electronically linkable records. At the same time, the sample survey paradigm underpinning data acquisition has evolved, and linkage of survey data to external data sources is now an important research tool in many areas of scientific research, with policy makers increasingly interested in using linked data to extend the scope and relevance of survey datasets. In particular, analysts now routinely use data linked from multiple sources to improve inference.

There are a variety of approaches to data integration based on probabilistic record linkage and statistical matching (Fellegi and Sunter, 1969; Elfeky *et al.*, 2002, 2003; Christen, 2012). Harron (2016) is a good introductory reference to the wider issues that arise with data linkage, while Harron *et al.* (2016) contains a number of articles that focus on the methodological and statistical issues that arise with data linkage. Here the focus is not on the various techniques of linkage and matching, but is primarily given to recent developments emphasizing inference procedures that allow for the probabilistic nature of linkage errors. This part is made up of two chapters, both of which are briefly summarized below.

Chapter 2: "On secondary analysis of datasets that cannot be linked without errors" (Zhang)

In the situation of linking two datasets A and B, linkage error occurs if (I) record a in A and b in B that correspond to two different units appear as a matched record ab in the linked dataset AB, or (II) record a in A and b in B that correspond to the same unit do not appear as a matched record ab in AB. Unless a unique and error-free identifier exists, or can be constructed in both datasets, and can be used to link them, linkage errors are by and large unavoidable. Adopting the viewpoint of a secondary analyst of linked data (i.e., an analyst who was not involved in the linking operation), this chapter discusses issues that arise if the aim is to carry out valid analysis, i.e., analysis that properly allows for the impact of linkage errors. To start, maximum likelihood estimation is considered in what DeGroot and Goel (1980) refer to as a "broken stick" situation, i.e., where complete lack of knowledge about the linkage process suggests that it is reasonable to assume that linkage errors are uniformly distributed across the space of possible links. Not surprisingly this approach fails in this case. Next, access to comparison data, i.e., scores related to the match probabilities of all possible links, is used to motivate unbiased estimation in a linear regression context. Unfortunately, this approach turns out to be somewhat impractical in large datasets. Finally, the possibility of restricting analysis to just those links that are *a priori* very probable (or are known to be correct) is examined. This approach is clearly very close to ignoring missingness when analysing survey data, and simple adjustments to the regression analysis that treats the linked data as correct are suggested.

Chapter 3: "Capture-recapture methods in the presence of linkage errors" (Di Consiglio *et al.*)

There is a large literature on population-size estimation, both in the area of official statistics as well as in the area of environmental statistics. In official statistics the traditional approach is to use some form of Dual System Estimation (DSE), with the same approach usually referred to as the Petersen estimator in environmental statistics. In any case, validity of these estimators depends on at least one of the data acquisition systems used to identify members of the population of interest being operated independently of the others. A basic assumption in both areas of application is then that there are no errors in identifying population units that have been "captured" by these independent systems. Since this identification is usually carried out by linking records in the different systems, this represents a linkage error scenario that differs from the one considered in the previous chapter. Here the concern is not only about errors when matching distinct population units in the same (overlap) sub-population, but also with errors in matching between overlap and non-overlap sub-populations. In particular, incorrect linking of overlap records tends to negatively bias the estimated population size, while missing overlap links, i.e., linking to non-overlap sub-populations tends to positively bias the estimated population size. Ding and Fienberg (1994) suggest an alternative to the standard DSE when there is linkage error. This chapter explores a number of extensions to this approach that do not require the asymmetry in linkage error probabilities that underpins the Ding and Fienberg approach.

Part 2: Statistical inference for fused data

Braverman (2008) describes statistical data fusion as the process of combining statistically heterogenous samples from marginal distributions in order to make inference about the unobserved joint distributions or functions of them. That is, statistical data fusion is the creation of joint distributional information from marginal data. Given two databases, say A and B, each containing data relating to different individuals in a population of interest, statistical data fusion then consists of the creation of a third database, made up of records from both A and B, that contains imputed values of database A variables for records originally from database B, and imputed values of database B variables for records originally from database A. Generally, it is assumed that both database A and database B contain the values of a set of common variables, i.e., variables defined the same way on both databases. The fused database then includes these common variables, with their common values. Data fusion is often referred to as statistical matching (in fact all three chapters summarized below use this nomenclature), and in this context is often confused with data linkage. However, a linked dataset represents a set of records that (in theory at least) actually exists for the population of individuals that provided the records making up databases A and B. In contrast, the fused dataset has no such basis in reality. Even if databases A and B correspond to the same population, there is no guarantee that there exist individuals in that population with the same values for the variables that appear in the fused dataset, which is essentially a synthetic construct that represents our best guess at realistic values of these variables that might be associated with individuals in this population. The data in a fused dataset can therefore be thought of as relating to a population of pseudo-individuals who are, in some sense, 'similar' to the real individuals who provided the information contained in databases A and B.

Let X denote the variables measured on A, Y denote the variables measured on B, and let Z denote the common variables measured on both A and B. A standard assumption in data fusion is then that X and Y are conditionally independent given Z in the population of interest. This Conditional Independence Assumption (CIA) justifies the creation of fused records by "matching" records in A and B on the basis of the values for Z. In this context, the matching does not have to be exact. For example, nearest neighbour fusion is carried out by matching records from A and B on the basis that their values of Z identify them as being "closest" with respect to an appropriate metric on the support of Z.

There is a close link between data fusion and missing data imputation (Little and Rubin, 1987), in the sense that the CIA is formally equivalent to the assumption that the values of Y (X) conceptually associated with the observed values of X (Y) in A (B) are missing at random given Z. Unlike the classical missing data problem, however, the 'missing data' in the case of data fusion do not always exist, since there is no requirement that the individuals defining database A overlap with the individuals defining database B (even if both A and B relate to the same underlying population). If one puts this concern to one side, however, it could be argued that data fusion is nothing more than missing data imputation. But the scale of missingness is extreme, particularly if A and B are large, and so missing data inference with fused data must be carried out with great care, particularly when imputation-based methods are used for fusion.

A key consideration with data fusion is the validity of the fused dataset, i.e., it is a feasible realisation from a joint distribution with the same marginal distributions as observed in A and B. How can we learn whether the fused data is feasible? Here we note that the Frechet-Hoeffding bounds for any joint distribution $F(x,y)$ in terms of its marginal distributions $F(x)$ and $F(y)$,

$$\max\{0, F(x) + F(y) - 1\} \leq F(x,y) \leq \min\{F(x), F(y)\}$$

can be used as a validity check for fused categorical data. Since these bounds hold for conditional (on Z) distributions, for discrete Y and X they become

$$\max\{0, \Pr(X = x|Z) + \Pr(Y = y|Z) - 1\} \leq \Pr(X = x, Y = y|Z)$$
$$\leq \min\{\Pr(X = x|Z), \Pr(Y = y|Z)\}.$$

As discussed in Chapters 4 and 6, these bounds, and refinements to them, can also be used to assess the uncertainty associated with any inference based on fused data.

Chapter 4: "An overview of uncertainty and estimation in statistical matching" (Scanu *et al.*)

This chapter develops the theory for data fusion under the CIA, and notes that the intervals defined by the Frechet-Hoeffding bounds constitute an uncertainty set for model parameters which always contains the CIA-based parameter estimates. The development in this chapter shows that this set can be made more compact by inclusion of extra-sample information, particularly in the form of logical constraints about parameter values. However, in general the chapter argues that the CIA assumption will usually not be appropriate, in which case one can use the Frechet-Hoeffding bounds to define a set of joint distributions for X and Y given Z, referred to as matching distributions, all of which are plausible. The chapter goes on to define a measure of the matching error, i.e., the difference between the matching distribution defined by the fused data and the actual, but unknown, joint distribution of X and Y given Z. Furthermore, this measure is bounded, with this bound itself estimable from the available data and therefore useful as a measure of uncertainty. The issue of estimation of this bound given that the fusion is carried out using complex sample survey data is also considered. Finally, the chapter explores the strong links between methods of ecological inference (Chambers and Steel, 2001; Wakefield, 2004) and data fusion, noting that some of the models used in the former could have the potential to improve methods of data fusion as well as analysis of fused data.

Chapter 5: "Auxiliary variable selection in a statistical matching problem" (D'Orazio *et al.*)

A critical issue in data fusion is how to define the variables in Z that are used to "glue" together records containing both X and Y data. In many applications there are a large number of variables that could be used to define Z, and it is not clear that all should be used. The key development in this chapter is how to choose variables for Z in order to make the uncertainty set defined by the Frechet-Hoeffding bounds as compact as possible. However, since these bounds get narrower as more variables are included in Z, a trade-off between narrower bounds and numbers of variables included in Z is necessary. This is accomplished by adding a penalty term to a measure of how wide the bounds are, where the penalty ensures that the contingency table defined by cross-tabulating the variables in Z does not contain too many cells with small counts.

Chapter 6: "Minimal inference from incomplete 2 x 2-tables" (Zhang and Chambers)

The final chapter in this part of this book takes a somewhat different tack when considering how inference should be carried out with data that have either been fused, or where missingness makes key parameters non-identifiable. As has already been noted, the CIA is often invoked in such cases. However, this is a very strong assumption, and one that the authors of this chapter wish to avoid. Instead, the focus turns to how one can integrate information from available marginal sources (which typically are of high quality) with the constraints provided by the Frechet-Hoeffding bounds. The key concept is one of corroboration, which is best explained via a thought experiment. Suppose both X and Y are binary and the objective of the data fusion exercise is to construct a large number of 2 x 2 tables corresponding to the levels of a categorical Z. The marginal probabilities $p_{1\bullet|z}$ that $X = 1$ and $p_{\bullet1|z}$ that $Y = 1$ for the table defined by $Z = z$ can be estimated separately from the X values in database A and the Y values in database B. As a consequence, we can estimate the Frechet-Hoeffding bounds for the joint probability $p_{11|z}$ that $X = 1$ and $Y = 1$ in this table. These are the limits of the interval

$$\max\left\{0, \hat{p}_{1\bullet|z} + \hat{p}_{\bullet1|z} - 1\right\} \leq p_{11|z} \leq \min\left\{\hat{p}_{1\bullet|z}, \hat{p}_{\bullet1|z}\right\}$$

where a "hat" denotes an estimated quantity. Now assume that we can make repeated independent draws from the sampling distributions of $\hat{p}_{1\bullet|Z}$ and $\hat{p}_{\bullet1|Z}$. Each draw generates a new set of estimated Frechet-Hoeffding bounds. If this process is carried out a large number of times then we can note that some potential values of $p_{11|Z}$ will occur more often than other values. These values are said to be more highly corroborated by the marginal information about X and Y derived from A and B respectively. The observed corroboration function is then defined as

$$c\left(p_{11|Z}\right) = \Pr\left(\max\left\{0, \hat{p}_{1\bullet|Z} + \hat{p}_{\bullet1|Z} - 1\right\} \leq p_{11|Z} \leq \min\left\{\hat{p}_{1\bullet|Z}, \hat{p}_{\bullet1|Z}\right\}\right)$$

where the probability on the right-hand side is defined with respect to the independent marginal sampling distributions of $\hat{p}_{1\bullet|Z}$ and $\hat{p}_{\bullet1|Z}$. The observed maximum corroboration set for $Z = z$ is the interval where $c\left(p_{11|Z=z}\right)$ is maximum, and a set of potential values for $p_{11|Z=z}$ is said to have a level of assurance equal to α if $c\left(p_{11|Z}\right) \geq \alpha$ on this set. This chapter explores the statistical properties of the observed corroboration function and statistical tests based on assurance levels in the data fusion case just described, as well as in the missing data situation where the tables cannot be completed because marginal information is missing.

Part 3: Population size modelling and estimation using multiple data sources

Combining data from multiple sources often reveals various over-coverage and under-coverage errors that reside in each source, as well as the problems associated with the delineation and classification of the statistical units of interest. The coverage problem is typically of special interest from the point of view of estimation of population size since the number of population units missed when combining two population registers is a key quality characteristic (under-coverage) as is the number of records erroneously included in the combined register (over-coverage).

Capture-recapture models for population-size estimation are well known and widely used. However, applications typically assume closed populations and perfect linkage of capture and recapture. The reality for human populations that are the target of most census-based applications is that the populations are open (i.e., dynamic) and that linkage is imperfect. Furthermore, the basic assumption of independence of capture and recapture is typically violated when these events are associated with identification in separate, incomplete, population registers. The following chapters address recent research aimed at the identification and fitting of models for population-size estimation that address this reality, as well as the design of data collection strategies for this purpose.

Chapter 7: "Dual and multiple system estimation with fully and partially observed covariates" (Van der Heijden *et al.*)

When the assumption of independence of capture by two separate registers cannot be justified, one possible approach is to introduce heterogeneous capture probabilities. That is, a model-based approach to specification of register inclusion probabilities is adopted, with these probabilities varying according to the levels of a specified set of categorical covariates. One such set of models are the log-linear models for the cross-classification of the register inclusion variables with these covariates. These models can be used to generate population-size estimates under the less restrictive assumption that captures are independent conditional on the covariates. However, the approach also requires that the same covariates be measured in all registers in the same way. This chapter develops this approach, as well as describing the necessary modifications when covariates that impact on register capture probability are only available from some of the registers, and also when the way in which a covariate is defined differs between registers.

Chapter 8: "Estimating population size in multiple record systems with uncertainty of state identification" (Di Cecco)

Like the previous chapter, this chapter is concerned with adjustments to standard methods of estimating the size of a finite population based on measuring the number of population units that are separately recorded on different registers. Here the issues addressed are where there is under-coverage (i.e., none of the registers "capture" the entire population), and also where the population inclusion information in the different registers are error-prone. That is, the observable counts that characterise the standard DSE approach to population-size estimation have errors. For example, in the case of two registers, say A and B, the counts corresponding to "Population unit included in both A and B", "Population unit only included in A", "Population unit only included in B" are all measured with error. The approach set out in this chapter is to characterise measurement error by a mixture model for the joint probability structure defined by the observed capture status of a register unit and a latent binary variable corresponding to whether the unit is actually a member of the population of interest or not. It turns out that a log-linear specification for this model can be fitted via maximum likelihood using the observed counts, and an estimate of the population size thereby obtained. Extensions to where capture probabilities are heterogeneous are also considered, and a Bayesian approach is developed.

Chapter 9: "Log-linear models of erroneous list data" (Zhang)

Traditional coverage estimation approaches readily allow for missing counts of the target population units but not erroneous counts of units that do not belong to the target population. A modelling approach that allows for over- as well as under-coverage errors in the available list enumerations can therefore be useful in a number of situations, such as population census, sizing of hard-to-reach and dynamic population, multiple screening procedure in health or other applications, etc. This chapter considers an important special case of the situation examined in the previous chapter, where multiple population registers are combined with a list derived from an independent coverage survey and where measurement errors (i.e., erroneous enumerations) exist in the registers but not in the coverage survey. Log-linear models for the observed enumerations are specified, which allow for under-coverage in the registers as well as the coverage survey, and with erroneous enumerations in the overlap between the registers and the coverage survey modelled via a pseudo-conditional independence assumption, i.e., where the probability of an erroneous enumeration in this overlap is the product of the marginal probabilities of erroneous enumeration in the different registers. Issues of model fitting are explored and applications to actual coverage analyses are described.

Chapter 10: "Sampling design and analysis using geo-referenced data" (Benedetti *et al.*)

A key advantage of data integration is the capacity to use the resulting dataset for new and more efficient inference. Integration of GPS data into a population register can lead to new insights about the population covered by the register, and also enable more powerful sample designs when register units are sampled. This chapter explores these issues by using simulation to compare the performances of modern approaches to spatial sampling that use the geo-coded information in the integrated data.

Geo-coding enables data across different sources to be matched on the basis of location. This allows data integration in situations where data linkage via natural population units such as person or business is not feasible. Combining multiple geo-coded datasets results in an enhanced population frame, which can improve the efficiency of sampling design, and increase the scope of cross-sectional and/or longitudinal statistical analysis. The discussion in this chapter is timely given the rapid uptake of geo-referenced data gathered from satellite surveillance as well as from GPS-equipped devices and from mobile and web networks. The chapter discusses two situations. In the first, a simulation study is used to illustrate the gains from integration of geo-coded data in the sample design of a coverage survey used to assess population register under-coverage rates for the nine provinces making up the Emilia Romagna region of Italy. Here the geo-coding is to census tract, and there are 35,585 census tracts in the region. In particular, the study examines the performances of sample weighted DSE estimates of register coverage rates for the nine provinces, based on application of modern spatial sampling methods to the selection of census tracts. The second situation considered is based on the fact that geo-referencing is not always error free. As a consequence, when these data are used to define a sampling design based on the integrated register, there is concern about how these locational errors could impact on estimation accuracy. This chapter addresses this issue via another simulation study, this time based on a business survey register containing geo-coded data. In this study different proportions of the units making up the register have their geo-locations suppressed, in the sense that these locations are changed to the locations of the centroids of sub-areas that include them. Again, a variety of modern spatial sampling methods are investigated, all dependent on this "contaminated" location information.

1.3 Summary

The nine chapters making up the rest of this book cover a wide variety of issues that arise when integrated register data are used for statistical purposes. These range from the impact of errors in data linkage and data fusion on statistical modelling of the register data to the impact of these errors on studies of register coverage to the impact of errors in the linked data (in particular geo-coding information) on statistical surveys that use the register as a sampling frame. What is striking about the developments presented in these chapters is that integration of data sources opens up many new opportunities for official statistics, but at the same time throws up a host of new problems arising from the inevitable imperfections of the integration process. Naively assuming that just because the integrated data are "Big" there are somehow no longer problems with analysis, and the integrated register can be used to produce population-based tabulations, is to live in a fool's paradise. It is certainly the case that sampling errors are smaller, and in many cases effectively zero, in such situations. But if one is aiming for anything approaching a valid statistical analysis, then allowance has to be made for errors due to deficiencies in the integration process. Doing so requires in turn that these errors be modelled in some appropriate way. That is, the "one size fits all" security of using sampling error to characterize inferential uncertainty that has been the dominant paradigm of survey-based official statistics for at least the last century has to be jettisoned. In its place, serious modelling work has to be carried out on how the process of integration has introduced uncertainty. In particular, official statisticians need to model and adjust for non-sampling error in their analyses. This book will hopefully provide some guidance in that respect.

References

ADRN: The UK Administrative Data Research Network. (2012). *Improving Access for Research and Policy*. Report from the Administrative Data Taskforce.

Braverman, A. (2008). Data fusion. In the *Encyclopaedia of Quantitative Risk Analysis and Assessment*, Volume 2. Wiley, New York.

Chambers, R.L. and Steel, D.G. (2001). Simple methods for ecological inference in 2 × 2 tables. *Journal of the Royal Statistical Society Series A* **164**, 175–192.

Christen, P. (2012). *Data Matching: Concepts and Techniques for Record Linkage, Entity Resolution, and Duplicate Detection*. Data-Centric Systems and Applications: Springer-Verlag Berlin Heidelberg.

DeGroot, M.H. and Goel, P.K. (1980). Estimation of the correlation coefficient from a broken random sample. *The Annals of Statistics*, **8**, 264–278.

Ding, Y. and Fienberg, S.E. (1994). Dual system estimation of Census undercount in the presence of matching error. *Survey Methodology*, **20**, 149–158.

Dunn, H. (1946). Record linkage. *American Journal of Public Health*, **36**, 1412–1416.

Elfeky, M.G., Verykios, V.S. and Elmagarmid, A.K. (2002). TAILOR: A record linkage toolbox. *Purdue e-Pubs*, Purdue University, Indiana.

Elfeky, M.G., Verykios, V.S., Elmagarmid, A.K., Ghanem, T.M. and Kuwait, W.A.R. (2003). Record linkage: A machine learning approach, a toolbox, and a digital government web service. *Purdue e-Pubs*, Purdue University.

Fellegi, I. and Sunter, A. (1969). A theory for record linkage. *Journal of the American Statistical Association*, **64**, 1183–1210.

Harron, K., Goldstein, H. and Dibben, C. (Editors) (2016). *Methodological Developments in Data Linkage*. Wiley: Chichester.

Harron, K. (2016). *An Introduction to Data Linkage*. Administrative Data Research Network Publication (eds. E. Mackay and M. Elliot), University of Essex: Colchester.

Little, R.J.A. and Rubin, D.B. (1987). *Statistical Analysis with Missing Data*. Wiley Series in Probability and Statistics. Wiley, New York, 1st edition.

Wakefield, J. (2004). Ecological inference for 2 x 2 tables (with Discussion). *Journal of the Royal Statistical Society Series A* **167**, 385–445.

2

On secondary analysis of datasets that cannot be linked without errors

Li-Chun Zhang

University of Southampton, Statistics Norway & University of Oslo

CONTENTS

2.1 Introduction

In record linkage (Fellegi and Sunter, 1969; Herzog et al., 2007), also known as entity or co-reference resolution (Christen, 2012), one aims to identify and bring together the records (with associated observations) in separate files, which correspond to the same entities (or units). Unless there exists a unique identifier that allows for exact matching, linkage errors are unavoidable when either the linked records do not actually refer to the same entity or if one fails to link the records that refer to the same entity.

TABLE 2.1

Summary of Census 2011 England and Patient Register linkage

Type	Pass	Number of Links	False Linkage Rate
Deterministic	1	30780660	0.00011
	2	11733197	0.00389
	3	1513471	0.00561
	4	2444838	0.00375
	5	1346432	0.00748
	6	121483	0.00886
	7	1007293	0.00100
	8	825069	0.01485
	9	35432	0.00100
Probabilistic	1	511239	0.02948
	2	298645	0.07165
Total			50617759

Table 2.1 provides a summary of the Census 2011 England and Patient Register (C-PR) linkage, carried out at the Office for National Statistics (ONS). See Owen et al. (2015) for some relevant aspects of the underlying methodology. Each of the eleven passes uses a different combination of the matching (or *key*) variables, which include name, birth date, sex, postcode, etc. In the end, about 3 million Census records are not linked to the Patient Register, whilst all the linked records are not true matches, as indicated by the estimated *false linkage rate (FLR)* at the different passes.

How can a *secondary analyst* conduct valid analysis based on the C-PR linkage data? This is the question we consider here. Two conditions help to clarify the secondary analyst's position. On the one hand, one may safely assume that the secondary analyst will not have access to all the key variables and the separate data files directly, nor the detailed knowledge or tools to replicate the actual linkage procedure. For instance, name, birth date and postcode are all used in the C-PR linkage, but these data are unlikely to be released to a secondary analyst for privacy protection. On the other hand, one must also assume that some *non-disclosive linkage comparison data* about how the records actually compare to each other can be made available. This is necessary to allow for any other possibility than to analyse the linked dataset as is; see, e.g., Gilbert et al. (2017) for a protocol of information exchange for data linkage. For instance, a linked pair of records can be assigned a score that reflects how well they agree with each other in terms of the key variables.

2.1.1 Related work

Different strands of approach can be discerned in the literature, when it comes to analysis of record linkage data that propagates uncertainty due to linkage errors. Unfortunately, none of the existing methods adequately addresses the question of secondary analysis posed above.

Nearly all the frequentist methods are based on the linkage model of the probability that a record in one dataset is linked to each of the records in another. Techniques such as regression analysis, estimation equation and analysis of contingency tables are studied by Scheuren and Winkler (1993, 1997), Lahiri and Larsen (2005), Chambers (2009), Chipperfield et al. (2011), Hof and Zwinderman (2012), Kim and Chambers (2012a, 2012b) and Chipperfield and Chambers (2015). Chipperfield and Chambers (2015) employ Monte Carlo simulation of both the agreement pattern between two datasets and the subsequent record linkage procedure to estimate the linkage model. The approach is restricted to someone who has access to all the key variables and the actual linkage procedure. Secondary analysts will therefore have to adopt greatly simplified assumptions, such as the exchangeable linkage error model of Chambers (2009), or resort to ad hoc approaches. For example, Hof and Zwinderman (2015) assume that each pair of records in the two datasets to be linked contribute independently to a pairwise pseudo-likelihood. But it is unclear how this pseudo-likelihood can yield valid inference.

For record linkage and analysis under the Bayesian framework, one may refer to Tancredi and Liseo (2011), Gutman et al. (2013), Sadinle (2014) and Stoerts et al. (2016), among others. Inference is based on the posterior distribution of the matched dataset. The modelling approach of the observed key variables, e.g., as described by Stoerts et al. (2016), builds on the hit-miss model of Copas and Hilton (1990). However, the distortion model is applied to the key variables, which are inaccessible to the secondary analyst, except in special circumstances. To hand out a large number of posterior match datasets to a secondary analyst may not always be practical, nor is it scalable to large datasets as in the C-PR case.

Goldstein et al. (2012) apply multiple imputation, where the units that are accepted as correctly linked have complete observations, while the units corresponding to the other records are considered to have missing observations. Gutman et al. (2015) take a different approach to multiply impute for a linked pair of records whether they are a true match or not, where the unlinked true matches are assumed to be missing at random. It should be noticed that these imputation methods are not constrained to the underlying linkage data structure. For instance, whether a linked pair of records is a true match or not affects whether other pairs of records can possibly be true matches or not. Restrictions like this are not built into these imputation methods.

2.1.2 Outline of investigation

We will investigate three different approaches in the sequels. We consider maximum likelihood estimation (MLE) of the parameter of interest in Section 2.3. This rests on the concept that the comparison data arises according to a probabilistic model, including distortions of the true values of the key variables that potentially cause linkage errors. However, as it will be demonstrated, the MLE based on models respecting 'the linkage data structure' (to be detailed in Section 2.2) is not straightforward in general. The discussion is related to the analysis of MLE of the correlation coefficient from a broken random sample of bivariate normal variables (DeGroot and Goel, 1980).

Next, using linear regression as the case-in-point, we consider unbiased regression estimator in Section 2.4, *conditional* on the available comparison data, and contrast it to the existing frequentist approach (e.g., Lahiri and Larsen, 2005) based on the so-called linkage model. We outline an approach to comparison data modelling that is applicable to common probabilistic and deterministic record linkage procedures, and discuss the scalability issue in the C-PR case. We find that valid analysis aimed at using *all* the records is exceedingly impractical for datasets of this magnitude.

The third approach consists of only making use of the links that have an almost-zero FLR, and adjusting the analysis based on the resulting *link subset* for the remaining false links and the missing true matches. While this clearly entails a loss of efficiency, it may be acceptable in situations like the C-PR linkage where one can still obtain a large accurate dataset, e.g., about 30 million records from the first pass where the estimated FLR is about 1 in 10 thousand. When all the links are true matches, the adjustment problem would become one of missing data. While the analysis of missing data is a challenging issue itself, there does exist a long tradition of using reweighting and imputation methods that are scalable to large datasets. We show in Section 2.5 how conditions for unbiased *link subset regression* can be developed, followed by simple asymptotic adjustments of the face-value regression analysis, similarly to the assumptions which allows one to base analysis on the observed data in the presence of missing data.

2.2   The linkage data structure

The linkage data structure underlying the separate datasets for record linkage sets the problem apart from those of missing data or measurement error. In the latter context, there is no ambiguity surrounding the units of analysis, but a value associated with a unit may be observed with an error or not observed at all. Often, it is reasonable to hypothesise that the random measurement errors or missing indicators are independent from one unit to another, or across some identified clusters of units. In record linkage, however, whether a record is linked to another, or which other record it is linked to, necessarily affects how the other records can possibly be linked, also when attention is restricted to within each block of linkable record sets. Different pairings of the records induce an uncertainty of the resulting analysis, even when all the associated outcome values are treated as fixed. A key issue is how to appropriately accommodate such inter-dependence that arises from the unit ambiguity rather than the random mechanism governing the outcome variables associated with the identified units. Any modelling approach that does not explicitly account for such inter-dependence due to the linkage data structure may be difficult to justify in general terms. Moreover, a related issue for implementation is how to deal with the combinatorics necessitated by such inter-dependence between two or more sets of records.

2.2.1 Definitions

We now provide a formal description of the linkage data structure underlying two separate datasets. The definition can easily be generalised to situations involving three or more datasets. Denote the separate datasets by A and B, of size n_A and n_B, respectively. Denote by $a \in A$ an arbitrary record from A, similarly for $b \in B$. Let i_a be an arbitrary fixed integer, indicating the *true* entity that $a \in A$ refers to. If $i_a = i_b$, for $a \in A$ and $b \in B$, then the pair of records (ab) is said to be a *match*, in which case a and b are both referred to as *matched* records between A and B. Whereas a is an *unmatched* record if $i_a \neq i_b$ for all $b \in B$. Similarly for b in B. We shall refer to

$$\Lambda_{AB} = \{i_a; a \in A\} \cup \{i_b; b \in B\}$$

as the *linkage data structure* between A and B. The linkage data structure is unknown except when there exists a unique identifier that allows for exact matching. The result of record linkage between A and B can be regarded as an estimated linkage data structure, denoted by $\hat{\Lambda}_{AB} = \{\hat{i}_a; a \in A\} \cup \{\hat{i}_b; b \in B\}$ where $\hat{i}_a = \hat{i}_b$ provided (ab) are a pair of linked records.

When there are only two datasets A and B, the linkage data structure can equally be given by the $n_A \times n_B$ *match matrix*, denoted by ω, where $\omega_{ij} = 1$ if the i-th record in A and the j-th record in B are a match, and $\omega_{ij} = 0$ otherwise. We shall assume that the duplicated records have been removed, so that $\sum_j \omega_{ij} \leq 1$ and $\sum_i \omega_{ij} \leq 1$. Denote by Ω the *match matrix space*, which contains all possible ω. Since each distinct ω corresponds to a distinct linkage data structure Λ_{AB}, the values of i_a and i_b being irrelevant otherwise, the space Ω defines effectively the space of linkage data structure.

Let ω_{ab} be the match indicator of an arbitrary pair of records (ab). Let $M(\omega) = \{(ab); \omega_{ab} = 1, a \in A, b \in B\}$ be the set of matches between A and B. Let $U_A(\omega)$ be the set of *unmatched* records in A and $U_B(\omega)$ that of B. In the absence of duplicated records, the sub-matrix of ω corresponding to M is a permutation matrix, which can be given by a permutation of the columns (or rows) of an identity matrix. This is a useful property of the match matrix representation. Let $\hat{\omega} = \omega_L$ be the *link matrix*, where $\omega_{L,ab} = 1$ if (ab) are a pair of linked records and $\omega_{L,ab} = 0$ otherwise. Let A_L consist of the linked records from A and B_L that from B. Record linkage can be regarded to yield estimates of M, U_A and U_B, where $\hat{M} = M(\omega_L) = \{(ab); \omega_{L,ab} = 1, a \in A_L, b \in B_L\}$, and $\hat{U}_A = A \setminus A_L$ and $\hat{U}_B = B \setminus B_L$, respectively.

We generally observe a sharp terminological distinction between link and match: while a match (ab) means $\omega_{ab} = 1$ and $i_a = i_b$ in truth, a link (ab) means $\hat{\omega}_{ab} = \omega_{L,ab} = 1$, which may or may not be a match.

Denote by g_{ab} the comparison data between $a \in A$ and $b \in B$, which is made available to the secondary analyst. Let G be the collection of available g_{ab}. For distinction, let g_a^* and g_b^* be the *full* comparison data accessible to the data provider who carries out the record linkage. In the C-PR linkage, g_a^* contains age, sex, name, birth date, postcode, etc. Not all of them will be released. Indeed, even if some of them are, such as age and sex, it is quite possible that they will not be specified as key variables.

As a possible example of available comparison data, g_{ab} can be a value which summarises how a compares to b, either absolutely or relatively. In particular, the *agreement indicator* is given by $g_{ab} = 1$ if a and b are considered to agree with each other, and $g_{ab} = 0$ otherwise. The link indicator, where $g_{ab} = 1$ if and only if $\omega_{L,ab} = 1$, may be considered a special agreement indicator that involves not only the comparison between a and b but also the decision whether or not to link them. For example, from a deterministic pass in Table 2.1, one can set the agreement indicator $g_{ab} = 1$ if a and b equal to each other in terms of all the relevant key variables, where $g_{ab} = 1$ and $g_{ab'} = 1$ may occur for different $b \neq b' \in B$. Whereas, if g_{ab} is a link indicator from the same pass, then one would typically not allow $g_{ab} = g_{ab'} = 1$ for $b \neq b'$, so that $g_{ab} = 1$ now means unique agreement.

As a scenario of *minimum* available comparison data, let the secondary analyst be only given the linked dataset, but not any of the unlinked records; G would then amount to the link matrix ω_L and nothing else.

2.2.2   Agreement partition of match space

Complete enumeration of the match space Ω gets quickly out of hand as the dataset sizes increase. To facilitate efficient partial enumeration, one can introduce a sensible partition of Ω, and only work with the sub-space where ω is most likely under some postulated statistical model. Below we describe an *agreement partition* of Ω, and consider some of its implications.

Without losing generality, let D be an agreement matrix defined by the secondary analyst, based on the available comparison data G, where $D_{ab} = 1$ if record $a \in A$ 'agrees' with $b \in B$, and $D_{ij} = 0$ otherwise. For instance, we have $D_{n \times n} = \omega_L$ provided minimum information, where $n = |A_L| = |B_L|$. Or, we can have $D_{n_A \times n_B} = \omega_L$ if the unlinked records in A and B are provided in addition. Or, provided the weights λ_{ab} from probabilistic record linkage under the Fellegi-Sunter (FS) paradigm (Fellegi and Sunter, 1969), one can define agreement to be the case if λ_{ab} exceeds some threshold value.

Let $\kappa = \sum_i \sum_j D_{ij}$ be the total no. agreements in D. For any $\omega \in \Omega$, let $d = \sum_a \sum_b D_{ab} \omega_{ab}$ be the no. *matched agreements* where $(\omega_{ab}, D_{ab}) = (1, 1)$, i.e., the no. pairs of record that are matches according to ω and in agreement with each other according to D. For example, provided $D_{n \times n}$ and one-one match so that each ω is a necessarily permutation matrix, we shall have in addition $n - d$ record pairs with $(\omega_{ab}, D_{ab}) = (1, 0)$, $\kappa - d$ with $(\omega_{ab}, D_{ab}) = (0, 1)$, and $n^2 - n - (\kappa - d)$ with $(\omega_{ab}, D_{ab}) = (0, 0)$.

Let Ω_0 be the *set* of ω that achieve the maximum value of d, denoted by d_0. The matrices in Ω_0 are referred to as the *most agreeable match matrices (MAMMs)*. For instance, under one-one linkage and one-one match, where D is also a permutation matrix and $\kappa = n$, there is only one MAMM with $d_0 = n$, i.e., $\omega = D$. More generally under one-one linkage, D can be an $n \times n$-matrix with $\kappa \geq n$. Insofar as D allows it, a one-one MAMM can have n elements with $(\omega_{ab}, D_{ab}) = (1, 1)$, 0 elements with $(\omega_{ab}, D_{ab}) = (1, 0)$, $\kappa - n$ elements with $(\omega_{ab}, D_{ab}) = (0, 1)$, and $n^2 - \kappa$ elements with $(\omega_{ab}, D_{ab}) = (0, 0)$.

Let Ω_1 contain all the ωs with the no. agreements given by $d_0 - 1$, as long as such matrices exist. This needs not to be always possible. For example, under one-one linkage and one-one match, the next highest possible no. agreements between D and ω is $n - 2$ but not $n - 1$, so that we have $\Omega_1 = \emptyset$. Let Ω_k be the set of match matrices with the no. agreements given by $d_0 - k$. The *agreement partition* of Ω is given by

$$\Omega(D) = \cup_{k=0}^{d_0} \Omega_k$$

Granted some confidence in the record linkage procedure, it seems natural that any modelling approach in practice would be such that the likelihood of ω is decreasing with k over the agreement partition of Ω. In fact, in the case of minimum comparison data, where $D = G = \omega_L$, there will be no information for a secondary analyst to place different likelihood on the ω's from the same Ω_k. Moreover, let $\omega \in \Omega_k$ and $\omega' \in \Omega_{k+1}$ be such that they differ only in one pair (ab), where $D_{ab} = 1$ and $\omega_{ab} = 1$ but $\omega'_{ab} = 0$. The likelihood ratio between the two based on D is given by

$$LR(\omega', \omega; D) = \frac{L(\omega'; D)}{L(\omega; D)} = \frac{\Pr(D | D_{ab} = 1; \omega')}{\Pr(D | D_{ab} = 1; \omega)} \cdot \frac{\Pr(D_{ab} = 1; \omega')}{\Pr(D_{ab} = 1; \omega)}$$

Now that ω and ω' coincide everywhere else except for (ab), a simplest modelling approach may be to assume

$$LR(\omega', \omega; D) = \frac{\Pr(D_{ab} = 1; \omega'_{ab} = 0)}{\Pr(D_{ab} = 1; \omega_{ab} = 1)}$$

Clearly, as a minimum feasibility requirement for secondary analysis, one must be provided certain linkage para-data so as to be able to model this ratio, in order to build an approach that respects the linkage data structure.

2.3   On maximum likelihood estimation

DeGroot and Goel (1980) consider MLE of the correlation coefficient of a bi-variate normal distribution, based on a broken sample. In other words, a random sample of n pairs is drawn from the distribution, but the observed data are only the first components of the n pairs and, separately, an unknown permutation of the second components of the n pairs, called a broken random sample from the distribution. Two approaches are studied, where the vector of permutation is either treated as fixed unknown parameters, or random variables that need to be integrated out of the observed likelihood.

They find that under the first approach the MLE is either the maximum or minimum sample correlation that can be calculated over all possible re-parings of the broken sample, which is unreasonable as a general method. Analogously, restricting ourselves to the situation of one-one match between A and B, this approach amounts to treating ω as parameters to be estimated, in addition to those of the analysis based on the linkage data. For instance, provided the variables involved in the analysis is independent of the errors of the key variables used for record linkage, the likelihood of ω can be factorised from that of the parameters of interest. Then, the discussion of agreement partition above suggests that the MLE of ω is unlikely to do well in the present context. For instance, in the case of $|\Omega_0| = 1$, a reasonable model of key variable errors may well imply the link matrix ω_L as the MLE of the true match matrix, from which no useful adjustment of the linkage errors will follow. Or, in the case of $|\Omega_0| > 1$, multiple MLEs are likely to be the case.

For the second approach, DeGroot and Goel (1980) consider the observed likelihood from integrating out the permutation vector, and a profile likelihood derived from it. Based on simulations in the case of $n = 5$, where explicit integration is feasible, they find the profile likelihood to be poor generally. Moreover, while the integrated likelihood looks very similar to the unbroken sample likelihood for many samples, the two can look very different for other samples. Still, this does not necessarily imply that the broken sample MLE is inconsistent asymptotically. To investigate this, however, one would need a way to obtain the MLE based on large broken samples, where explicit integration is impractical. The authors suggest that the MLE can in such cases be obtained via the EM algorithm, by treating the permutation vector as the missing data. But they did not examine the performance of this MLE.

Below we investigate this point in an example of regression analysis based on linkage data. Unfortunately, it is shown that the MLE may be biased and inconsistent, based on the missing information principle (Orchard and Woodbury, 1972) that underpins the EM algorithm.

Consider the setting of one-one match, under which $U_A = U_B = \emptyset$ and $n_A = n_B = n$, and ω is a permutation matrix such that $\omega \omega^T = \omega^T \omega = I$, where

I is the identity matrix. Consider the linear regression model

$$y_i = x_i^T \beta + \epsilon_i,$$

where ϵ_i is IID from $N(0, \sigma^2)$, and β is the p-vector of regression coefficients of interest. Let $X_A = X_{n_A \times k}$ be the covariates associated with A, and $(X_B, y_B) = (X_{n_B \times (p-k)}, y_{n_B \times 1})$ the remaining covariates and dependent variables associated with B, such that given the true match matrix ω, regression is based on $y_M = \omega y_B$ and partitioned design matrix $X_M = [X_A : \omega X_B]$.

The observed data consists of $z = (G, X_A, X_B, y_B)$, where G consists of the available comparison data. For instance, in the minimum case, we have $G = \omega_L$ provided one-one linkage. The complete data is (ω, z), containing in addition the unobserved ω. The MLE based on z needs to be obtained from integrating out ω from the distribution of the complete data. By the EM algorithm, one would maximise the expected complete-data log-likelihood conditional on the observed data. The complete-data log-likelihood and scores are given by

$$\ell_M(\beta, \sigma^2; \omega, z) = -\frac{n}{2} \log \sigma^2 - \frac{1}{2\sigma^2}(y_M - X_M^T \beta)^T (y_M - X_M^T \beta)$$

$$u_M(\beta; \omega, z) = \sigma^{-2}(X_M^T y_M - X_M^T X_M \beta)$$

$$= \sigma^{-2} \left(\begin{bmatrix} X_A^T \omega \\ X_B^T \end{bmatrix} y_B - \begin{bmatrix} X_A^T X_A & X_A^T \omega X_B \\ X_B^T \omega^T X_A & X_B^T X_B \end{bmatrix} \beta \right)$$

$$u_M(\sigma^2; \omega, z) = -n/\sigma^2 + (y_M - X_M \beta)^T (y_M - X_M \beta)/\sigma^4$$

For the present purpose, we now further simplify the situation and assume that the conditional distribution of ω given z is known, so that only β of interest needs to be estimated. It can easily be verified that the MLE $\hat{\beta}$ that maximises $E(\ell_M | z)$ is the same as the solution of $E[u(\beta; \omega, z) | z]$, which is given by

$$\hat{\beta} = E(X_M^T X_M | z)^{-1} E(X_M^T y_M | z)$$

$$= \begin{bmatrix} X_A^T X_A & X_A^T E(\omega | z) X_B \\ X_B^T E(\omega^T | z) X_A & X_B^T X_B \end{bmatrix}^{-1} \begin{bmatrix} X_A^T E(\omega | z) \\ X_B^T \end{bmatrix} y_B$$

Whereas the true complete-data MLE of β is given by

$$\hat{\beta}_M = (X_M^T X_M)^{-1}(X_M^T y_M) = \begin{bmatrix} X_A^T X_A & X_A^T \omega X_B \\ X_B^T \omega^T X_A & X_B^T X_B \end{bmatrix}^{-1} \begin{bmatrix} X_A^T \omega \\ X_B^T \end{bmatrix} y_B$$

such that $\hat{\beta}$ and $\hat{\beta}_M$ will almost surely differ, unless $\omega = E(\omega | z)$ as in the absence of linkage errors. Next, let $z_{(Y)} = (G, X_A, X_B)$, i.e., excluding y_B. Since

$E(y_M|\omega, z_{(Y)}) = X_M\beta$, we have

$$E(y_B|z_{(Y)}) = E\left(\omega^T E(y_M|\omega, z_{(Y)})|z_{(Y)}\right) = E(\omega^T X_M\beta|z_{(Y)})$$
$$= E(\omega^T[X_A : \omega X_B]\beta|z_{(Y)}) = [E(\omega^T|z_{(Y)})X_A : X_B]\beta$$

$$E(\hat{\beta}|z_{(Y)}) = \begin{bmatrix} X_A^T X_A^T & X_A^T E(\omega|z_{(Y)})X_B \\ X_B^T E(\omega^T|z_{(Y)})X_A & X_B^T X_B \end{bmatrix}^{-1}$$
$$\begin{bmatrix} X_A^T E(\omega|z_{(Y)})E(\omega^T|z_{(Y)})X_A & X_A^T E(\omega|z_{(Y)})X_B \\ X_B^T E(\omega^T|z_{(Y)})X_A & X_B^T X_B \end{bmatrix}\beta$$

It follows that the MLE $\hat{\beta}$ is biased conditional $z_{(Y)}$, unless

$$E(\omega|z_{(Y)})E(\omega^T|z_{(Y)}) = I,$$

which happens only if ω is uniquely determined given $z_{(Y)}$, such as in the absence of linkage errors. Moreover, since the bias does not vanish asymptotically as $n \to \infty$, neither is the MLE $\hat{\beta}$ consistent.

The difficulty of MLE above arises in situations where it is impractical to evaluate the observed likelihood directly, due to the large match matrix space Ω, as long as the modelling approach explicitly takes into account the linkage data structure. This does not mean that the MLE cannot be a reasonable estimator under some statistical model that provides a reasonable description of the variation in the observed data, say, (X_A, Y_B, ω_L), even though it does not strictly abide by the linkage data structure. What seems to be an open question at this stage is whether, or how, it is possible to develop a generally viable likelihood-based approach that respects the linkage data structure.

2.4   On analysis under the comparison data model

2.4.1   Linear regression under the linkage model

Linear regression has been the focus of many existing frequentist approach to analysis of linkage data. A central concept here is the linkage matrix. To focus on the idea, assume one-one match between A and B, and there are no unmatched entities at all. Suppose A contains all the covariates $X_{n \times p}$, and fix the ordering of the records in A, so that $X_M \equiv X$. Let y_L be the vector of dependent variables obtained from B via complete one-one record linkage, where the i-th component of y_L and the i-th row of X form a link. Let P be the unknown *linkage matrix*, so that $y_L = Py_M$, where the i-th component of y_M and the i-th row of X form a match, such that $\omega = I_{n \times n}$ by construction. Notice that P is not the observed link matrix ω_L. Provided *non-informative links*, by way of definition, we have, conditional on X,

$$E(y_L|X) = E[E(Py_M|G^*)|X] = E[E(P|G^*)E(y_M|X)|X]$$
$$= E[E(P|G^*)|X]E(y_M|X) = E(P|X)X\beta = QX\beta$$

where G^* is the comparison data accessible to the data linker, which may contain part of X. The face-value least squares estimator $(X^TX)^{-1}X^Ty_L$ is biased since $(X^TX)^{-1}(X^TQX) \neq I$, unless $Q = I$. Lahiri and Larsen (2005) propose the adjusted unbiased estimator

$$\hat{\beta}_{LL} = (X^TQ^TQX)^{-1}X^TQ^Ty_L$$

Kim and Chambers (2012a, 2012b) and Hof and Zwinderman (2012) propose extensions under specific assumptions to allow more than two datasets and sample-register linkage. Chambers (2009) notes despite that it is in principle possible to use the optimal $(X^TQ^T\Sigma_L^{-1}QX)^{-1}X^TQ^T\Sigma_L^{-1}y_L$, where $\Sigma_L = Cov(y_L, y_L|X)$, the gains can be small, because the covariance matrix Σ_L needs to be estimated and is risk-prone to misspecifications.

The assumption of non-informative links above means that conditional on X, the distortions of the key variables are independent of the dependent variable y. It can seem reasonable in many applications (Lahiri and Larsen, 2005). For the C-PR linkage, this would be the case if the distribution of the other key variables (e.g., name, birth date) is independent of the study variables (e.g., health), given X that e.g., for example, may include sex and age. Many large undertakings use similar record linkage methods; see, e.g., the Australian Census Longitudinal Dataset (Zhang and Campbell, 2012). Informative linkage models remain currently an open question (Chambers and Kim, 2015).

The matrix $Q = E(P|X)$ contains the pairwise true (q_{ii}) and false (q_{ij}) link probabilities, over both the distortion process of G^* and the record linkage procedure. However, what one generally needs for analysis is the distribution $f(P|X)$, e.g., in order to evaluate the variance of $\hat{\beta}_{LL}$. This distribution cannot be derived from Q except in the following special case. Let $A = \{a_1, a_2\}$ and $B = \{b_1, b_2\}$. Under complete one-one linkage, there are only two possible linkage matrices, i.e., $P = I_{2\times2}$ or $P = J_{2\times2} - I$, where J is the unity matrix, so that $q_{11} = q_{22}$ must be the probability of $P = I = \omega$ and $q_{12} = q_{21}$ that of $P = J_{2\times2} - I$. Otherwise, $f(P|X)$ is inaccessible to a secondary analyst.

Chambers (2009) proposes secondary analysis under the simplest linkage model, where $q_{ii} = \lambda$ and $q_{ij} = (1 - \lambda)/(n - 1)$ for $j \neq i$, which is referred to as the *exchangeable linkage error (ELE)* model. This has the advantage of allowing the secondary analyst to obtain the matrix Q based on a single parameter related to the linkage errors. Despite its simplicity and suitability to secondary analysis, it is unclear what $f(P|X; \lambda)$ is under the ELE model, without which the scope of secondary analysis is somewhat limited.

2.4.2 Linear regression under the comparison data model

We now consider an approach under the comparison data model $f(G|\omega)$. The modelling of $f(G|\omega)$ itself will be discussed in the two following sections. As above let X be from A and y from B, and assume complete one-one match, so that $X = X_M$, $y_M = \omega y$ and $y = \omega^T y_M$. We assume that ω is uniformly distributed over Ω, provided the records are randomly arranged in A and B. The same assumption is used by DeGroot and Goel (1980). Assume *non-informative* comparison data, i.e.,

$$f(G|\omega, X, Y; \xi) = f(G|\omega; \xi)$$

where ξ is the vector of parameters of the *comparison data model*. Then,

$$E(y|G, X) = E(\omega^T|G)E(y_M|X) = Q_G X\beta$$

where $Q_G = E(\omega^T|G)$ is calculated over

$$f(\omega|G; \xi) = f(G|\omega; \xi) / \sum_{\omega \in \Omega} f(G|\omega; \xi) \tag{2.1}$$

since $f(\omega)$ cancels out in the numerator and denominator. In this way, the conditional distribution $f(\omega|G)$ can be obtained from the comparison data model $f(G|\omega; \xi)$, so that inference can be made conditional to how the records actually compare to each other in a given situation.

In the case of linear regression analysis, and similarly to $\hat{\beta}_{LL}$ above, a conditionally unbiased estimator of β given (G, X) can be given as

$$\hat{\beta}_G = (X^T Q_G^T Q_G X)^{-1} X^T Q_G^T y = Hy \tag{2.2}$$

provided $(X^T Q_G^T Q_G X)^{-1}$ exists, where $H = (X^T Q_G^T Q_G X)^{-1} X^T Q_G^T$. For the variance-covariance matrix $V(\hat{\beta}_G)$, we have

$$V(\hat{\beta}_G|X) = H(\Delta_G + \sigma^2 Q_G Q_G^T)H^T$$

where $V[E(y|y_M)|G, X] = \sigma^2 Q_G Q_G^T$, and $\Delta_G = E[V(y|y_M)|G, X]$ with the (ij)-th element: $\Delta_{G,ij} = \beta^T X^T \tau_{ij} X\beta + \sigma^2 \text{Trace}(\tau_{ij})$, for $\tau_{ij} = E(\omega_i \omega_j^T|G) - E(\omega_i|G)E(\omega_j^T|G)$ and ω_i is the i-th column of ω and ω_j similarly. To estimate σ^2, consider

$$\hat{e} = y - \hat{y}_M = \omega^T y_M - \hat{y}_M = \omega^T y_M - X\hat{\beta}_G = (I - XH)\omega^T y_M$$
$$E(\hat{e}^T \hat{e}|G) = \sigma^2 \text{Trace}(D) + \beta^T X^T DX\beta$$

where $D = E(\omega R \omega^T|G, X)$ and $R = (I - XH)^T(I - XH)$. A plug-in estimate of σ^2 can be given on substituting $\hat{\beta}_G$ for β.

From the point of view of secondary analysis, the use of $f(\omega|G)$ under the comparison data model is the key difference to the previous approach under the linkage model. It is worth emphasising that both models can provide valid analysis, but are suitable on *different* premises. If the data linker releases the linkage matrix distribution $f(P|X)$ together with the link dataset, then the secondary analysts can of course develop the analysis of interest under the linkage model. In reality though, data providers have so far been unable to provide even the Q-matrix, beyond the simple ELE model, let alone $f(P|X)$. Insofar as one is only provided the linkage comparison data G, comparison data modelling provides a principled option.

2.4.3 Comparison data modelling (I)

Consider first an approach to comparison data modelling, which can be related to probabilistic record linkage under the FS-paradigm (Fellegi and Sunter, 1969). Without losing generality, let $g_a^* = h$ be the full comparison data of $a \in A$ and $g_b^* = j$ that of $b \in B$, for $h, j = 1, ..., \mathcal{K}$. As in Stoerts et al. (2015), we assume that potential distortions of the comparison data are independent between the entities, so that

$$f(G^*|\omega) = \prod_{(ab) \in M} f(g_a^*, g_b^*|\omega_{ab} = 1) \prod_{a \in U_A} f(g_a^*|a \in U_A) \prod_{b \in U_B} f(g_b^*|b \in U_B)$$

Notice that it is in principle possible for the data linker to provide these probabilities as the comparison data, without disclosing the key variables. When only comparison data $G = \{g_{ab}; a \in A_L, b \in B_L\}$ is available, which is the mostly likely situation in practice, the comparison data model can be obtained from integrating out g_a^* and g_b^* in the full comparison data model above, which yields

$$f(G|\omega) = \prod_{(ab):\omega_{ab}=1} f(g_{ab}|\omega_{ab} = 1) \tag{2.3}$$

To model the relevant probabilities, consider the hit-miss distortion model proposed by Copas and Hilton (1990). Envisage a binary trial that results in a *miss* with probability α, following which the observed value h is randomly distributed with probability p_h, where p_h is the frequency of the true value h. Let $\mathcal{K} = n_A + n_B - m$ be the number of entities in A and B, where $m = |M|$. Assume distinct full comparison data of all the entities, so that $p_h \equiv 1/\mathcal{K}$. For a simple comparison data model, let g_{ab} be the agreement indicator. We have

$$\mu_{ab} = f(g_{ab} = 1|\omega_{ab} = 1) = (1 - \alpha)^2 + 2\alpha(1 - \alpha)/\mathcal{K} + \alpha^2/\mathcal{K}$$
$$\mu_{ab}^c = f(g_{ab} = 0|\omega_{ab} = 1) = 1 - \mu_{ab} = \left(1 - (1 - \alpha)^2\right)(1 - 1/\mathcal{K})$$

which may be referred to as the *exchangeable agreement (EA)* model. The secondary analyst needs only to be provided the estimates of α and $\mathcal{K}$ (or m).

It seems intuitive that the EA model can result in the ELE model under suitable conditions. For instance, assume one-one complete match, where

$h, j = 1, ..., n$ and $U_A = U_B = \emptyset$. Let $\mu_{ab} = \mu(\alpha)$ under the EA model, where g_{ab} is the agreement indicator. Suppose record linkage (denoted by ϕ) proceeds as follows: link (ab) if they are in unique agreement; link randomly among the records that are not in unique agreement, subjected to the one-one linkage constraint. This would result in the ELE model, with the parameter $\lambda(\alpha, \phi) \equiv q_{ii} = f(P_{ii} = 1 | \omega_{ii} = 1)$, where $\omega = I_{n \times n}$ by definition.

TABLE 2.2
Ratio μ/λ under one-one linkage given miss rate α and size n

α	$n = 3$	$n = 10$	$n = 50$	$n = 300$	$n = 1000$
0.02	0.974/0.987	0.965/0.981	0.961/0.975	0.961/0.964	0.960/0.962
0.05	0.935/0.966	0.912/0.950	0.904/0.925	0.903/0.910	0.903/0.907
0.10	0.873/0.937	0.829/0.895	0.814/0.846	0.811/0.827	0.810/0.824
0.20	0.760/0.878	0.676/0.790	0.647/0.712	0.641/0.691	0.640/0.688

Whilst $\mu(\alpha)$ has a closed expression, $\lambda(\alpha, \phi)$ can only be obtained via Monte Carlo simulation of G and the subsequent linkage procedure ϕ. Table 2.2 provides the ratio between μ and λ given α and n, where each λ is based on 5000 simulations. Clearly, $\mu/\lambda < 1$ in all the cases. At the one end, observe that $\mu/\lambda \to 1$ as $n \to \infty$ for any fixed α, because the chance of agreement following a miss tends to zero, as well as that of match by random linkage. At the other end, μ/λ is further away from 1 for smaller n and larger α, where the effect of one-one linkage restriction is intuitively the strongest.

TABLE 2.3
Examples of $Q_G^T = E(\omega | G)$ given α and $n = 3$.

$G = G_1$			$Q_G^T : \alpha = 0.02$			$Q_G^T : \alpha = 0.1$		
1	0	0	0.998	0.001	0.001	0.96	0.02	0.02
0	0	1	0.001	0.001	0.998	0.02	0.02	0.96
0	1	0	0.001	0.998	0.001	0.02	0.96	0.02
$G = G_2$			$Q_G^T : \alpha = 0.02$			$Q_G^T : \alpha = 0.1$		
1	0	0	0.97	0.01	0.01	0.86	0.07	0.07
0	1	1	0.00	0.50	0.50	0.02	0.49	0.49
0	0	0	0.02	0.49	0.49	0.12	0.44	0.44

For a very simple illustration of linear regression under the comparison model (2.3), suppose one-one complete match with $n = 3$. The matrix Q_G for $\hat{\beta}_G$ by (2.2) depends on the observed G. Two examples are given in Table 2.3. At $\alpha = 0.02$, the probability $f(G_1 | \omega)$ is 0.923 for observing G_1, given which $\hat{\beta}_G$ is almost as efficient as knowing ω directly, since $Q_G \approx I$. At $\alpha = 0.1$, the probability for observing G_1 drops to 0.666. The regression is less efficient given $G = G_2$, as can be seen from the corresponding Q_G. The difference is rooted in $f(G | \omega) = \mu^d (1 - \mu)^{n-d}$, where $d = \sum_{(ab)} \omega_{ab} g_{ab}$ is the no. observed agreements among the matches in $M(\omega)$. Let $d_0(G)$ be the maximum possible

d given G, where $d_0(G_1) = 3$ and $d_0(G_2) = 2$. We have

$$\sum_{\omega \in \Omega} f(G|\omega) = \sum_{d=0}^{d_0(G)} m_d f_d$$

where $f_d = f(G|\omega)$ is the same for any ω with the same d, and m_d the no. such ω's. The likelihood ratio is $f_{d+1}/f_d = \mu/(1 - \mu)$. The agreement partition of Ω induced by G_1 is $m_d = (2, 3, 1)$ for $d = (0, 1, 3)$, and the likelihood of $\omega = G_1$ is f_3, which is $\mu^2/(1 - \mu)^2$ times that of the next highest f_1. Whereas, given G_2, we have $m_d = (2, 2, 2)$ for $d = (0, 1, 2)$, where there are two ω's with the highest likelihood f_2, which is $\mu/(1 - \mu)$ times that of the next highest f_1. The uncertainty is clearly greater on observing $G = G_2$ than $G = G_1$.

2.4.4 Comparison data modelling (II)

Multi-pass deterministic linkage is common in practice. Let $L_k = A_k \times B_k$ contain the n_k links from the kth pass, for $k = 1, ..., K$. These are the unique pairs of records in complete agreement with each other at the k-th pass. The unlinked records are left to the subsequent passes. Let A_0 contain the unlinked records in A and B_0 those of B. The linkage results in the partition

$$A \times B = \bigcup_{k=0}^{K} \bigcup_{j=0}^{K} A_k \times B_j$$

Let $g_{ab} = (k, j)$ indicate which domain of the partition the record pair (ab) belongs to, the collection of which constitute the comparison data G.

For any $(ab) \in L_k$ and $1 \leq k \leq K$, let $\pi_k = f(\omega_{ab} = 1|g_{ab} = (k, k))$, so that $1 - \pi_k$ is the FLR in L_k (e.g., Table 2.1). We have

$$\mu_{ab} = f(g_{ab} = (k, k)|\omega_{ab} = 1) = \pi_k f(g_{ab} = (k, k))/f(\omega_{ab} = 1)$$

One can estimate μ_{ab} by $\pi_k n_k/m$. Ideally, one would like the data provider to be able to provide $\mu_{ab} = f(g_{ab} = (k, j)|\omega_{ab} = 1)$ for all the other combinations (k, j) as well. However, this can be demanding in practice. As a possible simplification, let $L = \bigcup_{k=1}^{K} L_k$ contain all the linked pairs, and let

$$\mu_{ab} = f((ab) \notin L|\omega_{ab} = 1) = 1 - \sum_{k=1}^{K} f(g_{ab} = (k, k)|\omega_{ab} = 1)$$

which assumes a uniform distribution of g_{ab} outside of the linked pairs.

The above comparison model is of course a great simplification of the reality because the matches are unlikely to be uniformly distributed over $R = (A \times B) \setminus L$. Indeed, among all the unlinked pairs of records, one can expect $m_R = m - \sum_{k=1}^{K} \pi_k n_k$ matches. One can explore the bounds of μ_{ab} in the different parts of R. For instance, let R_k be the subset of R involving either $a \in A_k$ or $b \in B_k$, where one can expect at most $2n_k(1 - \pi_k)$ matches, so that $f\big((ab) \in R_k | \omega_{ab} = 1\big) \leq 2(1 - \pi_k)n_k/m$, which provides an upper bound of μ_{ab} over R_k. Moreover, let $R_0 = A_0 \times B_0$, where one can expect at least $m_R - 2\sum_{k=1}^{K}(1 - \pi_k)n_k$ matches, so that a lower bound of μ_{ab} over R_0 is $f\big((ab) \in R_0 | \omega_{ab} = 1\big) \geq 1 - \sum_{k=1}^{K}(2 - \pi_k)n_k/m$.

Let us consider the C-PR case for an illustration. Let n_A be the size of the Census dataset, and n_B that of the Patient Register. Let n_L be the number of links (Table 2.1), and m that of matched entities. For easy numerical appreciation, suppose $m = 51$, $n_A = 53$ and $n_B = 55$, all in millions.

- For the k-th deterministic pass, where $k = 1, ..., 9$, the FLR $1 - \pi_k$ is given in Table 2.1. We set $\mu_{ab} = \pi_k n_k/m$ for $(ab) \in L_k$.

- We do not have other information than Table 2.1 about the two probabilistic passes, denoted by L_{10} and L_{11}, respectively. We therefore treat them as if they had arisen from deterministic linkage.

- We calculate also the upper-bound probability $\tilde{\mu}_{ab}$ in R_k, where $1 \leq k \leq K$, and the lower-bound probability $\tilde{\mu}_{ab}$ in R_0.

Table 2.4 summarises all the probabilities calculated based on Table 2.1.

TABLE 2.4
Probabilities and bounds for the CPR data

k	Domain	$\mu_{ab:k}$	Domain	Bound $\tilde{\mu}_{ab:k}$
1	L_1	0.6035	R_1	0.0001
2	L_2	0.2292	R_2	0.0018
3	L_3	0.0295	R_3	0.0003
4	L_4	0.0478	R_4	0.0004
5	L_5	0.0262	R_5	0.0004
6	L_6	0.0024	R_6	0.0000
7	L_7	0.0197	R_7	0.0000
8	L_8	0.0159	R_8	0.0005
9	L_9	0.0007	R_9	0.0000
10	L_{10}	0.0097	R_{10}	0.0006
11	L_{11}	0.0054	R_{11}	0.0008
0	$\notin L$	0.0100	R_0	0.0050

Scalability is an important issue when it comes to large population datasets like in the C-PR case. Consider the generic task of evaluating

$$E(t_\omega | G) = \sum_\Omega t_\omega f(\omega | G) = \sum_\Omega t_\omega f(G | \omega) / \sum_\Omega f(G | \omega)$$

where t_ω is a function of ω, such as $t_\omega = \omega^T$ for Q_G in (2.2). Complete enumeration of Ω is infeasible in most situations. Let a sample from Ω be

$s = \{\omega_1, ..., \omega_{n_s}\} \subset \Omega$. If the ωs are sampled randomly, a Hájek-type estimate of $E(t_\omega|G)$ is $\bar{t}_H = \sum_{\omega \in s} t_\omega f(G|\omega)/\sum_{\omega \in s} f(G|\omega)$. Whereas if $\omega \sim f(\omega|G)$ for $\omega \in s$, a Monte Carlo estimate is $\bar{t} = \sum_{\omega \in s} t_\omega/n_s$. The Metropolis-Hastings algorithm provides a general approach to the latter: at every step a new ω' is proposed from the current ω, where the proposal is symmetric and the acceptance rate is $\min\{1, f(\omega'|G)/f(\omega|G)\}$.

We can analyse the acceptance rate for the C-PR linkage data directly under the comparison data model above. Let $\omega = \omega_L$ be the match matrix that agrees exactly with the link matrix, where $\omega_{ab} = 1$ if and only if $(ab) \in L$. Some least departures from ω_L are depicted generically in the following sub-matrices of $\{a, a', a''\} \times \{b, b', b''\}$:

$$[\omega_L] = \begin{bmatrix} 1 & 0 & 0 \\ 0 & 1 & 0 \\ 0 & 0 & 0 \end{bmatrix} \quad [\omega_1] = \begin{bmatrix} 1 & 0 & 0 \\ 0 & 1 & 0 \\ 0 & 0 & 1 \end{bmatrix}$$

$$[\omega_2] = \begin{bmatrix} 0 & 0 & 1 \\ 0 & 1 & 0 \\ 0 & 0 & 0 \end{bmatrix} \quad [\omega_3] = \begin{bmatrix} 0 & 1 & 0 \\ 1 & 0 & 0 \\ 0 & 0 & 0 \end{bmatrix}$$

where $(ab) \in L_k$, $(a'b') \in L_{k'}$, for $1 \le k \ne k' \le 11$, and $(a''b'') \in R_0$. The first row in each matrix is the sub-vector $(\omega_{ab}, \omega_{ab'}, \omega_{ab''})$, similarly for the second and third rows.

- The only difference between ω_L and ω_1 is that ω_1 contains an extra match $(a''b'')$ in R_0. We have $f(\omega_1|G)/f(\omega_L|G) = \mu_{ab:0} = 0.0100$, or at least 0.0055 according to Table 2.4.

- According to ω_2, a pair in R_k is actually a match instead of a pair in L_k. We have $f(\omega_2|G)/f(\omega_L|G) = f\big((ab'') \in R_k|\omega_{ab''} = 1\big)/\mu_{ab:k} \le \bar{\mu}_{ab:k}/\mu_{ab:k}$, which is close to zero for all $1 \le k \le 11$ in Table 2.4.

- Given ω_3, (ab) in L_k and $(a'b')$ in $L_{k'}$ are false links, while matches (ab') in R_k and $(a'b)$ in $R_{k'}$ fail to be linked. We have $f(\omega_3|G)/f(\omega_L|G) \le \bar{\mu}_{ab:k}\bar{\mu}_{ab:k'}/\mu_{ab:k}\mu_{ab:k'}$, which is almost zero for all $1 \le k \ne k' \le 11$.

The acceptance rates are thus too low for the Metropolis-Hastings algorithm to be numerically efficient. Given $m > n_L$, the match matrices that have ω_L as the sub-matrix for the linked records, plus $m - n_L$ additional matches among $A_0 \times B_0$, have all the same likelihood. The number of such matrices is

$$\big((n_A - n_L)!/(n_A - m)!\big)\big((n_B - n_L)!/(n_B - m)!\big)/(m - n_L)!$$

Any resampling method for $f(\omega|G)$ will be exceedingly impractical when the outcome space is of such a magnitude.

2.5   On link subset analysis

2.5.1   Non-informative balanced selection

The scalability problem above arises because one aims to utilise all the matched entities that exist between the separate datasets. However, the likelihood of ω can be flat over an extremely large space due to the missed matches. In the C-PR case, it seems therefore natural to ask whether it is possible to limit the analysis to the 50 million linked records, or perhaps even a subset of it. Using again linear regression as the case-in-point, we study below the conditions for valid *link subset analysis*.

To start with, assume one-one match between A and B. Let (A_1, A_2) be a bipartition of A and (B_1, B_2) that of B, where (A_1, B_1) form a *complete match space (CMS)*, which means that none of the records A_1 has match to a record $b \notin B_1$. Moreover, (A_2, B_2) form a CMS, provided (A_1, B_1) form a CMS. Let X_1 and X_2 be the regression design matrices associated with A_1 and A_2, respectively. Let y_1 and y_2 be the vectors of dependent variables associated with B_1 and B_2, respectively. For given (X_1, X_2), denote the true vectors of dependent variables by $y_{M1} = \omega_1 y_1$ and $y_{M2} = \omega_2 y_2$, respectively. The true match matrix between A and B is the block diagonal matrix, where ω_1 and ω_2 are the blocks on the diagonal. Let G_1 and G_2 be the respective comparison data. We say that (A_1, B_1) form a *non-informative complete selection (NICS)* of (A, B), if (A_1, B_1) form a CMS and the comparison data G_1 is non-informative. Likewise for NICS (A_2, B_2).

Lemma 1 *Let (A_1, A_2) be a bipartition of A and (B_1, B_2) that of B, where (A_1, B_1) and (A_2, B_2) form NICS of (A, B). Let $\hat{\beta}_G$, $\hat{\beta}_{G1}$ and $\hat{\beta}_{G2}$ be given by (2.2) based on (X, y), (X_1, y_1) and (X_2, y_2), respectively. Let $T_1 = X_1^T Q_1^T Q_1 X_1$ and $T_2 = X_2^T Q_2^T Q_2 X_2$. Let $r = T_1^{-1} T_2$ and $\phi = (I + r)^{-1} r$. We have*

$$\hat{\beta}_G = (I - \phi)\hat{\beta}_{G1} + (I - \phi)r\hat{\beta}_{G2} = \hat{\beta}_{G1} + [(I - \phi)r\hat{\beta}_{G2} - \phi\hat{\beta}_{G2}]$$

Proof: By virtue of NICS, we obtain $T = (QX)^T(QX) = T_1 + T_2$, where $Q = E(\omega^T | G)$ for $y_M = \omega y$, $Q_1 = E(\omega_1^T | G)$ for $y_{M1} = \omega_1 y_1$ and $Q_2 = E(\omega_2^T | G)$ for $y_{M2} = \omega_2 y_2$. Similarly, we have $(QX)^T y = X_1^T Q_1^T y_1 + X_2^T Q_2^T y_2$. Using $T^{-1} = T_1^{-1} - (I + r)^{-1} r T_1^{-1} = (I - \phi) T_1^{-1}$, we obtain

$$\hat{\beta}_G = (I - \phi) T_1^{-1} (X_1^T Q_1^T y_1 + T_2 T_2^{-1} X_2^T Q_2^T y_2) = (I - \phi)\hat{\beta}_{G1} + (I - \phi)r\hat{\beta}_{G2} \quad \square$$

Notice that, while $\hat{\beta}_G$ is not simply a convex combination of $\hat{\beta}_{G1}$ and $\hat{\beta}_{G2}$, we do have $(I - \phi) + (I - \phi)r = I$, since

$$(I - \phi)r - \phi = [I - \phi - (I + r)^{-1}]r = [I - (I + r)^{-1}(r + I)]r = 0$$

Thus, Lemma 1 shows that regression is valid based on a link subset, provided it is a part of some NICS-partition of all the records. Of course, unless the FLR is zero, one cannot be sure that the linked records, denoted by (A_L, B_L), form a NICS of (A, B). But it is possible to consider record linkage as a means for identifying a probable NICS. Let π_L be the probability that (A_L, B_L) forms a NICS. We have then $E(\hat{\beta}_{GL}|G, X) = \beta$ with probability π_L, where $\hat{\beta}_{GL}$ is given by (2.2) as if (A_L, B_L) form a NICS. Moreover, when (A_L, B_L) does not form a NICS, let $\omega_{(L)}$ be the $n_A \times n_L$ sub-matrix of the true match matrix $\omega_{n_A \times n_B}$, consisting of the columns of ω that correspond to B_L. Provided $\omega_{(L)}^T 1 = 1$, i.e., all the records in B_L do have matches in A even though they are not all in A_L, the true regression design matrix of y_L is given as $X_{ML} = \omega_{(L)}^T X$. Lemma 2 below shows that the bias of $\hat{\beta}_{GL}$ is then limited by the probability that (A_L, B_L) does not form a CMS, i.e., $1 - \pi_L$.

Lemma 2 *Let* $\hat{\beta}_{GL} = T_L^{-1}(X_L Q_L^T y_L)$, *where* $T_L = X_L^T Q_L^T Q_L X_L$ *and* Q_L *is calculated as if* (A_L, B_L) *form a NICS. Let* $f(\Omega_L|G)$ *be the probability that* (A_L, B_L) *forms a NICS. Provided* $\omega_{(L)}^T 1 = 1$, *where* $\omega_{(L)}$ *is the true match sub-matrix of* B_L, *we have*

$$E(\hat{\beta}_{GL}|G, X) = \pi_L \beta + (1 - \pi_L) T_L^{-1} X_L^T Q_L^T E(\omega_{(L)}^T|G) X \beta$$

Notice that, in the situation where not all the records in B_L have matches in A, the expectation of $\hat{\beta}_{GL}$ cannot be given without making some additional assumptions about the unmatchable y-values in B_L. It is nevertheless clear that in creating the link set (A_L, B_L) it helps to increase the probability that (A_L, B_L) forms a CMS. Whilst it can be difficult to evaluate the probability π_L for large datasets, it seems intuitive that limiting the FLR would help the course, since we necessarily have $\pi_L = 1$ for any link set if the FLR is zero.

The results above concern the adjusted $\hat{\beta}_G$ (2.2) applied to a link subset of records. For large population datasets, it would be much more practical if it is possible to employ the face-value estimator

$$\hat{\beta}_L = (X_L^T X_L)^{-1}(X_L^T y_L)$$

in a more direct manner. Building on the reasoning behind Lemma 1 and 2, we now develop an asymptotic result for $\hat{\beta}_L$ based on large datasets.

Let (A_{ML}, A_{cL}) be a bipartition of A_L and (B_{ML}, B_{cL}) that of B_L, where A_{ML} contains the n_M records that are correctly linked to B_{ML}, where $n_M \le n_L$, and A_{cL} the $n_c = n_L - n_M$ records that are incorrectly linked to B_{cL}. Let (X_{ML}, X_{cL}) be associated with (A_{ML}, A_{cL}), and (y_{ML}, y_{cL}) with (B_{ML}, B_{cL}). Let $T_{ML} = X_{ML}^T X_{ML}$, $T_{cL} = X_{cL}^T X_{cL}$, $r_L = T_{ML}^{-1} T_{cL}$ and $\phi_L = (I + r_L)^{-1} r_L$. Even without knowing which records are in A_{ML} and B_{ML}, we have

$$\hat{\beta}_L = (X_{ML}^T X_{ML} + X_{cL}^T X_{cL})^{-1}(X_{ML}^T y_{ML} + X_{cL}^T y_{cL})$$
$$= (I - \phi_L) T_{ML}^{-1} X_{ML}^T y_{ML} + (I - \phi_L) r_L T_{cL}^{-1} X_{cL}^T y_{cL} \tag{2.4}$$

Assume the existence of X_{McL} for y_{cL}, where $E(y_{cL}|X_{McL}) = X_{McL}\beta$, despite the records in B_{cL} are incorrectly linked. With large datasets and low FLR in mind, we put down the following conditions. Asymptotically, as $n_L \to \infty$, (A_L, B_L) is said to form a *non-informative balanced selection (NIBS)* if

(B1) $n_M/n_L \overset{P}{\to} \pi_M$, where $\pi_M \in (0, 1]$

(B2) $n_M^{-1} X_{ML}^T X_{ML} \overset{P}{\to} \mu_x \mu_x^T + \Sigma_x$ and $n_c^{-1} X_{cL}^T X_{cL} \overset{P}{\to} \mu_x \mu_x^T + \Sigma_x$

(B3) $n_c^{-1} X_{cL}^T X_{McL} \overset{P}{\to} \mu_x \mu_x^T$

These are e.g., the case if X_{ML}, X_{cL} and X_{McL} are random subsets of X.

Theorem 1 *Provided (B1) - (B3), asymptotically as $n_L \to \infty$, we have*

$$E(\hat{\beta}_L) \overset{P}{\to} \beta - (1 - \pi_M)(I - \psi)\beta$$
$$n_L V(\hat{\beta}_L) \overset{P}{\to} \sigma^2 \Delta_x^{-1}$$

where $\psi = (\mu_x \mu_x^T + \Sigma_x)^{-1} \mu_x \mu_x^T$ and $\Delta_x = \mu_x \mu_x^T + \Sigma_x$.

Proof: From (B1) and (B2), we have $r_L \overset{P}{\to} I(1 - \pi_M)/\pi_M$ and $I - \phi_L \overset{P}{\to} \pi_M I$. For the first term of (2.4), it follows that

$$(I - \phi_L) T_{ML}^{-1} X_{ML} y_{ML} \overset{P}{\to} \pi_M \beta = \beta - (1 - \pi_M)\beta$$

For the second term of (2.4), it follows from (B3) that $T_{cL}^{-1} X_{cL}^T X_{McL} \overset{P}{\to} \psi$, such that $(I - \phi_L) r_L T_{cL}^{-1} X_{cL}^T y_{cL} \overset{P}{\to} (1 - \pi_M)\psi\beta$. Combining the two terms yields the first result. For the variance of $\hat{\beta}_L$, we observe

$$V(\hat{\beta}_L) = (I - \phi_L) V(\hat{\beta}_{ML})(I - \phi_L)^T + (I - \phi_L) r_L T_{cL}^{-1} V(X_{cL}^T y_{cL}) T_{cL}^{-1} r_L^T (I - \phi_L)^T$$

where $\hat{\beta}_{ML} = T_{ML}^{-1} X_{ML}^T y_{ML}$. $\square$

Notice that $V(\hat{\beta}_L)$ is not specified in terms of $(X_{ML}, X_{cL}, X_{McL})$ since these are not observed. The asymptotic bias of using the as is NIBS link subset is thus bounded by the FLR $1 - \pi_M$, provided which an asymptotic bias-adjusted estimator of β is given as

$$\hat{\beta}_A = \left(I - (1 - \pi_M)(I - \hat{\psi}) \right)^{-1} \hat{\beta}_L \tag{2.5}$$

The estimator (2.5) is simple to calculate, which requires only the face-value $\hat{\beta}_L$ by (2.4) and the estimated mean and covariance matrix of X_L.

2.5.2 Illustration for the C-PR data

The results above suggest a practical approach to the C-PR data (Table 2.1), which consists of only making the links that have almost-zero false linkage probabilities, and adjusting the analysis based on the resulting link subset, i.e., of all the links that otherwise could have been made. When all the records in this link subset are true matches, they necessarily form a CMS, and the adjustment problem would become one of missing data. Developing general approaches to adjustment of such link subset is beyond the scope here. Below we consider some broad alternatives, and the implied costs and benefits both for the data provider and the secondary analysts.

Consider linear regression $y_i = x_i^T \beta + \epsilon_i$, where y_i and x_i reside separately in the two datasets, and $\beta_{p \times 1}$ is the vector of regression coefficients. Suppose $Cov(\epsilon_i, \epsilon_j) = \sigma^2$ if $i = j$ and 0 if $i \neq j$. For our purpose here, suppose the x_i is standardised, where $\mu_x = E(x_i) = 0$ and $\Sigma_x = V(x_i)$, so that $\psi = 0$ and $V(\hat{\beta}) = O(1/n_L)$ in Theorem 1, provided NIBS link subset.

It can be seen in Table 2.1 that over 97% of the linked records have FLR below 0.01, where the exceptions are the deterministic pass 8 and the two probabilistic passes. Notice that the two probabilistic passes are the most costly for the data provider, while being associated with the largest FLRs.

Consider L_1, consisting of the 30 million links from the first deterministic pass, where the FLR is as low as 0.00011. The secondary analyst may (i) assume NICS and obtain $\hat{\beta}_{GL1}$ given by (2.2) based on L_1, as under Lemma 1, or (ii) assume NIBS and obtain $\hat{\beta}_{A1} = \hat{\beta}_{L1}/\pi_1$ by (2.5) under Theorem 1, where $\psi = 0$ and $\hat{\beta}_{L1}$ is the face-value OLS (2.4) based on L_1 and $1 - \pi_1 = 0.00011$. Clearly, the NICS assumption is not the case given the detected FLR, so that option (i) is not entirely justified and, as discussed in Section 2.4.4, the computation of $Q_1 = E(\omega_1^T|G)$ for L_1 is impractical. In contrast, option (ii) is straightforward and sidesteps the computation related to the conditional distribution $f(\omega|G)$. Provided the NIBS assumption, the plausibility of both $\hat{\beta}_{L1}$ and $\hat{\beta}_{A1}$ seem intuitive given the low FLR.

Similar reasoning holds for all the other link subsets $L_2, ..., L_{11}$ separately. Consider now combining the subset estimates $\hat{\beta}_{Ak} = \hat{\beta}_{Lk}/\pi_k$. Let $s \subseteq \{1, 2, ..., 11\}$ be an index set. A combined estimator based on the link subsets in s given by

$$\hat{\beta}_{As} = \sum_{k \in s} w_k \hat{\beta}_{Ak} \quad \text{and} \quad V(\hat{\beta}_{As}) = \sum_{k \in s} w_k^2 V(\hat{\beta}_{Ak})$$

where $1/w_k = v_k W$, and $W = \sum_{k \in s} 1/v_k$, and v_k are constants of choice. A natural choice of v_k is $v_k = V(\hat{\beta}_{Ak})$, provided the NIBS assumption holds in each subset L_k for $k \in s$. Another possibility is simply $v_k = 1/n_k$, so that $w_k = n_k/n_s$ where $n_s = \sum_{k \in s} n_k$. This is somewhat less efficient than $v_k = V(\hat{\beta}_{Ak})$ if the NIBS assumption actually holds in all the subsets. But it remains plausible in case the NIBS assumption holds for the union of the subsets $L_s = \bigcup_{k \in s} L_k$ but not separately in each subset.

TABLE 2.5
Illustration of Census and Patient Register link subset regression

Link Subsets	Combined FLR	Order of Standard Error
$\{1\}$	1.1×10^{-4}	1.80×10^{-4}
$\{1, 2, 3, 4, 5, 6, 7, 9\}$	16.1×10^{-4}	1.43×10^{-4}
$\{1, 2, 3, 4, 5, 6, 7, 8, 9, 10, 11\}$	25.2×10^{-4}	1.41×10^{-4}

Table 2.5 shows some summary statistics of three alternative settings of s. The combined FLR is given by $1 - \sum_{k \in s} n_k \pi_k / n_s$, and the order of standard error (SE) is $O(1/\sqrt{n_s})$. The first alternative uses only L_1, provided the NIBS assumption. The second alternative uses the 8 subsets that have FLRs below 0.01. The combined FLR is 15 times of that of L_1 alone, while the SE is potentially reduced by about 20%. This alternative may be approximately valid if the NIBS assumption holds reasonably for the union of the relevant subsets, provided $v_k = 1/n_k$. It does require additional work for the data provider. All the linked records are used for the final alternative. The combined FLR is about 50% higher than the previous alternative, while the potential reduction of SE is minimal and may not be worth the heightened risk of bias. As noted before, the two probabilistic record linkage procedures are much more resource demanding compared to the direct deterministic passes.

In short, analysis based on link subsets with near-zero FLRs have practical advantages in terms of the resources required, for the secondary analysts as well as the data provider. It seems that the necessary adjustment for the potential linkage error bias may be less complicated, and may carry a lower risk due to the misspecification of comparison data model.

2.6   Concluding remarks

We have considered three broad approaches regarding how secondary users can carry out analysis of datasets that cannot be linked without errors, provided non-disclosive comparison data between the relevant records.

It is still an open question at this stage regarding the MLE under models that respect the linkage data structure. For an analysis that aims at using all the matched entities in the separate datasets, there are two key methodological challenges. Firstly, more accurate comparison data models may be needed, e.g., to accommodate heterogeneous distortion errors of the key variables, especially for datasets of small to moderate sizes. Next, the computation related to the conditional distribution $f(\omega|G)$ seems unscalable at the moment.

For analysis of large population datasets, link subset analysis may offer a practical alternative. The result obtained above, provided asymptotic NIBS link subset, needs to be extended in order to handle non-uniform or potentially informative selection of the link subset.

It is important to observe the key role of the data provider that emerges from our investigation, which is to prepare the datasets and to disseminate appropriate linkage comparison data, in order for the secondary analyst to have the possibility to carry out valid statistical analysis, and to account for the uncertainty that arises from not knowing the true matched entities.

Bibliography

[1] CHAMBERS, R.C. (2009). Regression Analysis of Probability Linked Data. *Statisphere*, Vol. 4: http://ro.uow.edu.au/eispapers/762/

[2] CHAMBERS, R.C. and KIM, G. (2015). Secondary Analysis of Linked Data. In *Methodological Developments in Data Linkage (eds. K. Harron, H. Goldstein and C. Dibben)*, Chapter 5.

[3] CHIPPERFIELD, J.O., BISHOP, G.R. and CAMPELL, P. (2011). Maximum likelihood estimation for contingency tables and logistic regression with incorrectly linked data. *Survey Methodology*, **37**, 13–24.

[4] CHIPPERFIELD, J.O. and CHAMBERS, R.C. (2015). Using bootstrap to account for linkage errors when analysing probabilistically linked categorical data. *Journal of Official Statistics*, **31**, 397–414.

[5] CHRISTEN, P. (2012). A survey of indexing techniques for scalable record linkage and deduplication. *ISEE Transactions on Knowledge and Data Engineering*, **24**.

[6] COPAS, J.B. and HILTON, F.J. (1990). Record linkage: Statistical models from matching computer records. *Journal of the Royal Statistical Society, Ser. A*, **153**, 287–320.

[7] DEGROOT, M.H. and GOEL, P.K. (1980). Estimation of the correlation coefficient from a broken random sample. *The Annals of Statistics*, **8**, 264–278.

[8] FELLEGI, I.P. and SUNTER, A.B. (1969). A theory for record linkage. *Journal of the American Statistical Association*, **64**, 1183–1210.

[9] GILBERT, R., LAFFERTY, R., HAGGER-JOHNSON, G., HARRON, K., ZHANG, L-C., SMITH, P., ... GOLDSTEIN, H. (2017). GUILD: GUidance for Information about Linking Data sets. *Journal of Public Health*, DOI : 10.1093/pubmed/fdx037

[10] GOLDSTEIN, H., HARRON, K. and WADE, A. (2012). The analysis of record-linked data using multiple imputation with data value priors. *Statistics in Medicine*, **31**, 3481–3493.

[11] GUTMAN, R., AFENDULIS, C.C. and ZASLAVSKY, A.M. (2013). A Bayesian procedure for file linking to analyze end-of-life medical costs. *Journal of the American Statistical Association*, **108**, 34–47.

[12] GUTMAN, R., SAMMARTINO, C.J., GREEN, T.C. and MONTAGUE, B.T. (2015). Error adjustments for file linking methods using encrypted unique client identifier (eUCI) with application to recently released prisoners who are HIV+. *Statistics in Medicine*, **35**, 115–129.

[13] HERZOG, T.N., SCHEUREN, F.J. and WINKLER, W.E. (2007). *Data Quality and Record Linkage Techniques*. Springer.

[14] HOF, M.H.P. and ZWINDERMAN, A.H. (2012). Methods for analysing data from probabilistic linkage strategies based on partially identifying variables. *Statistics in Medicine*, **31**, 4231–4242.

[15] HOF, M.H.P. and ZWINDERMAN, A.H. (2015). A mixture model for the analysis of data derived from record linkage. Statistics in Medicine, **34**, 74–92.

[16] KIM, G. and CHAMBERS, R.C. (2012a). Regression analysis under incomplete linkage. *Computational Statistics and Data Analysis*, **56**, 2756–2770.

[17] KIM, G. and CHAMBERS, R.C. (2012b). Regression analysis under probabilistic multi-linkage. *Statistica Neerlandica*, **66**, 64–79.

[18] LAHIRI, P. and LARSEN, M.D. (2005). Regression analysis with linked data. *Journal of the American Statistical Association*, **100**, 222–230.

[19] ORCHARD, T. and WOODBURY, M.A. (1972). A missing information principle: theory and applications. *Proc. Sixth Berkeley Symp. on Math. Statist. and Prob.*, Vol. 1, 697–715.

[20] OWEN, A., JONES, P. and RALPHS, M. (2015). Large-scale Linkage for Total Populations in Official Statistics. In *Methodological Developments in Data Linkage (eds. K. Harron, H. Goldstein and C. Dibben)*, Chapter 8.

[21] SADINLE, M. (2014). Detecting duplicates in a homicide registry using a Bayesian partitioning approach. *Annals of Applied Statistics*, **8**, 2404–2434.

[22] SCHEUREN, F. and WINKLER, W. E. (1993). Regression analysis of data files that are computer matched. *Survey Methodology*, **19**, 39–58.

[23] SCHEUREN, F. and WINKLER, W. E. (1997). Regression analysis of data files that are computer matched – Part II. *Survey Methodology*, **23**, 157–165.

[24] STOERTS, R., HALL, R. and FIENBERG, S. (2016). A Bayesian approach to graphical record linkage and de-duplication. *Journal of the American Statistical Association*, **111**, 1660–1672.

[25] TANCREDI, A. and LISEO, B. (2013). A hierarchical Bayesian approach to record linkage and population size problems. *The Annals of Applied Statistics*, **5**, 1553–1585.

[26] ZHANG, G. and CAMPBELL, P. (2012). Data Survey: Developing the Statistical Longitudinal Census Dataset and identifying its potential uses. *Australian Economic Review*, **45**, 125–133.

3

Capture-recapture methods in the presence of linkage errors

Loredana Di Consiglio

Istat, National Institute of Statistics, Rome, Italy

Tiziana Tuoto

Istat, National Institute of Statistics, Rome, Italy

Li-Chun Zhang

University of Southampton, Statistics Norway & University of Oslo

CONTENTS

3.1   Introduction

A standard approach to population-size estimation is the capture-recapture method originated by [45] and [37], and reviewed in [14], [43], [46] and [52], [53]. Applications to human population events date back to 1949 [54]. The method is applied to human populations for evaluating, for instance, the size of the undercount in censuses (see [63], [64] , [24], [27], [30]), the number of people affected by specific diseases or using illegal drugs [4], the number of victims in civil wars [3]. See [7] for a recent collection of capture-recapture methods in the Social and Medical Sciences. A traditional focus of study has been the assumption of homogeneity or independence of the recaptures. Extensions to account for heterogeneity or list dependency can be found in, e.g., [40], [6], [10], [67]. See, e.g., [20] for a Bayesian approach.

Recently, the use of capture-recapture method has received increasing attention in connection with producing census-like population statistics based on multiple administrative registers, instead of the traditional census. This has highlighted two other methodological challenges. Firstly, the administrative registers may contain considerable erroneous enumerations (of out-of-scope individuals) in addition to under-counts; see, e.g., Chapters 8 and 9. Secondly, in many applications there does not exist a unique identifier that can be used to link all the administrative registers, in which case linkage errors are unavoidable, and the commonly made assumption of perfect matching is violated. In this chapter, we concentrate on this problem and study different proposals that adjust the capture-recapture estimation by explicitly taking into account the linkage errors.

The chapter is organized as follow: Section 2 briefly recounts the capture-recapture model and its assumptions, introducing notations used in the rest of the chapter. Section 3 reviews some common frequentist and Bayesian record linkage approaches and estimation of the associated of linkage errors. Sections 4 and 5 present correction methods for the capture-recapture estimation with two or more lists. Examples are included both to illustrate the effect of linkage errors on the bias of population estimates and to demonstrate the gains of the proposed adjustments. Finally, Section 6 provides some concluding remarks and directions for future works.

3.2   The capture-recapture model: short formalization and notation

The capture-recapture method for estimating the unknown size of a population (N) requires one to compare two (or more) enumeration lists of the target population units, in order to count the number of units 'captured' in each list, as well as those that are 'recaptured' in any combination of the lists.

TABLE 3.1
Contingency table of the counts in two lists

		List 2	
		Present	*Absent*
List 1	*Present*	n_{11}	n_{10}
	Absent	n_{01}	n_{00}

TABLE 3.2
Contingency table of the counts based on three lists

	List 1			
	Present		*Absent*	
	List 3		List 3	
List 2	Present	Absent	Present	Absent
Present	n_{111}	n_{110}	n_{011}	n_{010}
Absent	n_{101}	n_{100}	n_{001}	n_{000}

Take the example of two lists. Let n_{1+} and n_{+1} be the counts of population units in the first and second lists, respectively. Let n_{11} be that in both lists, so that $n_{10} = n_{1+} - n_{11}$ is the number of units only in List 1 and $n_{01} = n_{+1} - n_{11}$ that only in List 2. The counts can be organized in a 2x2 contingency table 3.1, where n_{00} is the (unobservable) number of units missed by both lists.

The so-called dual system estimator (DSE) of the population size N is given by

$$\tilde{N}_P = n_{1+} \times n_{+1}/n_{11}. \tag{3.1}$$

The DSE can be motivated in various ways. See, e.g., [63] for a discussion of the relevant assumptions. Of particular relevance to our discussions later, notice that, provided the probability of being enumerated in List 1 is constant for all the population units, an estimate of this homogenous (i.e., constant) catch probability is given by

$$\tilde{\pi}_{1,P} = n_{11}/n_{+1} \tag{3.2}$$

Similarly, provided homogeneity of capture in List 2, we have

$$\tilde{\pi}_{2,P} = n_{11}/n_{1+} \tag{3.3}$$

A more general approach, which includes the DSE as a special case, can be based on log-linear models of the K-list capture-recapture data [22]. In particular, due to the structural zero cell count, i.e., of the units missing from all the lists, one must set the interaction terms involving all the K factors to zero. Fienberg [22] shows that the maximum likelihood estimate (MLE) of the unobserved cell count can then be given in a simple generic expression. For the 2-list situation, this becomes

$$\tilde{n}_{00} = n_{10} \times n_{01}/n_{11}. \tag{3.4}$$

It follows that the MLE of N is given by $\tilde{N}_{ML} = n + \tilde{n}_{00} = \tilde{N}_P$, where $n = n_{11} + n_{10} + n_{01}$. Thus, let the 3-list capture-recapture counts be arranged as in Table 3.2, denoted by n_{ijk}, for $i, j, k = 0, 1$. The MLE is then given by $\tilde{N}_{ML} = n + \tilde{n}_{000}$, where

$$\tilde{n}_{000} = \frac{n_{111} n_{001} n_{100} n_{010}}{n_{101} n_{011} n_{101}} \tag{3.5}$$

3.3 The linkage models and the linkage errors

A key step of the capture-recapture method is thus the identification of the common units in two (or more) enumeration lists, which is referred to as record linkage. In this section we describe the probabilistic record linkage frameworks, mainly in order to formalize the definition and the estimation of linkage errors.

3.3.1 The Fellegi and Sunter linkage model

A frequentist theory for record linkage is given in the seminal paper by Fellegi and Sunter [21]. We refer to Herzog et al. [29] for a comprehensive account. Given two lists, say L_1 and L_2, of size N_1 and N_2, let $\Omega = \{(a,b), a \in L_1 \text{ and } b \in L_2\}$ be the complete set of all possible pairs, of size $|\Omega| = N_1 \times N_2$. Record linkage between L_1 and L_2 is viewed as a classification problem, by which the pairs in Ω are assigned to two subsets, M and U, independent and mutually exclusive, such that:

M is the link set (a=b)

U is the non-link set (a≠b).

Common identifiers (the linking variables) are chosen and, for each pair, a comparison function is applied in order to obtain the corresponding comparison vector, denoted by γ. Let r be the ratio between the conditional probability of γ given that the pair belongs to set M and the conditional probability of γ given that the pair belongs to set U. It is the likelihood ratio test statistic for H_0: $(a,b) \in M$ against H_1: $(a,b) \in U$, i.e.,

$$r = \frac{P(\gamma | (a,b) \in M)}{P(\gamma | (a,b) \in U)} = \frac{m(\gamma)}{u(\gamma)} \tag{3.6}$$

Hence, those pairs for which r is greater than the upper threshold value T_m are assigned to the set of linked pairs, M^*; those pairs for which r is smaller than the lower threshold value T_u are assigned to the set of unlinked pairs, U^*; if r falls in the range (T_u, T_m), no decision is made automatically and the pair is classified by clerical review.

The thresholds are chosen to minimize false link probability, denoted by β, and false non-link probability, denoted by $1 - \alpha$, which are defined as follows:

$$\beta = \sum_{\gamma \in \Gamma} u(\gamma) P(M^* | \gamma) = \sum_{\gamma \in \Gamma_{M^*}} u(\gamma) \qquad \text{where} \quad \Gamma_{M^*} = \{\gamma : T_m \leq m(\gamma)/u(\gamma)\}$$

(3.7)

$$1 - \alpha = \sum_{\gamma \in \Gamma} m(\gamma) P(U^* | \gamma) = \sum_{\gamma \in \Gamma_{U^*}} m(\gamma) \qquad \text{where} \quad \Gamma_{U^*} = \{\gamma : T_u \geq m(\gamma)/u(\gamma)\}.$$

(3.8)

In applications, the probabilities m and u can be estimated by treating the true link status as a latent variable, and using the EM algorithm [33]. Alternatively, [34] applies Bayesian latent class and Bayesian log linear models to fit the mixture models [36].

For situations with more than two files, there exist a number of efforts for simultaneous record linkage; see [50], [49], [55], [62] , [25]. However, making these methods scalable for "industrial strength" applications to population datasets is still an open question at the moment. A plausible approach is to perform pairwise linkage of the involved lists. However, a key difficulty is that separate pairwise linking of lists does not guarantee the transitivity of the linkage decisions. For example, suppose in the 3-list problem, the record a in List 1 is linked to the b in List 2 based on the 2-list approach above, and b is linked to the record c in List 3 based on separate 2-list linkage, but a and c are not linked based on separate 2-list linkage. This can happen because each 2-list linkage has its own acceptance threshold. There are now 5 possible 'decisions': a, b, and c refer to the same individual, a and b refer to the same individual but c refers to another one, a and c refer to the same individual but b refers to another one, b and c refer to the same individual but a refers to another one, all a, b, and c refer to different individuals. However, there is no logical way to reduce this 5-category classification decision to the three 2-category classification decisions in a transitive manner.

Thus, for record linkage of multiple files, a widespread practice is first to create a master frame. Each of the rest lists is then separately linked to the master frame one-by-one. For instance, one could start by linking List 1 and List 2, and then link the resulting frame with List 3. Such a procedure has the advantage of requiring only two linkage operations, without the need to solve discrepancies, while the pairwise approach requires three linkage operations, with the need to solve discrepancies as explained. Accordingly to the described two-step process, let $1 - \alpha_1$ be the probability of missing a link in the first operation and $1 - \alpha_2$ that of the second operation; moreover, let β_1 be the probability of false link in the first operation and β_2 that of the second operation. We assume this approach in Section 5.

3.3.2   Definition and estimation of linkage errors

The two types of linkage error, namely false links and missed matches, play a
central role in adjusting statistical analysis based on linkage data, including
the capture-recapture estimators. However, as it is well known in practice,
although the decision-rule approach of Fellegi and Sunter [21] is very effective for link identification, the formulae (3.7) and (3.8) are generally less reliable for the evaluation of the errors β and $1 - \alpha$. Notice that, conceptually, by
(3.7) and (3.8) the parameters β and $1 - \alpha$ are defined for each element of the
cross-product comparisons $\Omega = L_1 \times L_2$, i.e., the probability of a pair of records
ending in M^* or U^* given it is in U or M. Clearly, this definition does not take
into account the inter-dependence inherent of any linkage procedure, where
(for general cases treated here) a record in L_1 may be linked to not more than
one record in L_2 and vice versa. Thus, for statistical analysis of linkage data,
one should avoid using the estimates of β and $1 - \alpha$, which are produced
to facilitate the decision rule, and instead conduct post-linkage evaluation
which can accommodate the actual linkage inter-dependence. Tuoto [61] proposes a supervised learning method to predict both types of linkage error,
without relying on strong distribution assumptions, as in [5]. Alternatively,
[11] apply a bootstrap method to the actual linkage procedure to evaluate
the mismatch probabilities, where the effects of linkage inter-independence
are replicated in the bootstrap linkage results.

   Next, putting aside the issue of estimation, it may be convenient or necessary to adopt alternative definitions of the linkage errors than those applied
in (3.7) and (3.8). Consider for instance the DSE in (3.1), let n_{11}^* be the count
of linked records between Lists 1 and 2, which relates to the count of true
matches n_{11} via two numbers: the number of missed matches among the n_{11}
true matches, and the number of false links among the n_{11}^* actual links.

   In practice, the probability of missed matches can be defined using the
denominator $(N_1 - n_{11}^*) + (N_2 - n_{11}^*)$, which is the number of unlinked records
in either List 1 or 2, instead of the size of U^*, i.e., among all the links that
are not made. Notice that U^* is a much larger set than $(N_1 - n_{11}^*) + (N_2 - n_{11}^*)$.
Moreover, it is worth noting that one can expect the number of false links
involving the actually linked records to be much lower than the number of
false links between unlinked records, because the former implies two linkage
errors simultaneously, i.e., missing the true match and erroneously linking
the matched record to a different record.

   It should be pointed out that the *false link rate* defined in this way is a different quantity to the *false match rate* used to adjust regression analysis (e.g.,
in [8]), where the latter is defined in relation to the number of actual links
n_{11}^*. While both target the *same number* of false links among the links made,
the two rates are not the same because they have different denominators.

   Finally, it is worthwhile to note also the different relevance the two kinds
of error may have. Take the DSE (3.1), it is clear that n_{11}^* should be adjusted
upwards due to the missing matches, but downwards due to the false links

among the actual links. Depending on the balance between the two errors, the face-value DSE using n_{11}^* instead of n_{11} could be either an over- or under-estimate of the population size. Notice that, in practice, it is often easier to reduce the false link rate, e.g., by using more restrictive acceptance criteria. However, this would inevitably increase the number of missing matches. In many studies of animal populations, based on the recognition of individual animals from natural markings (e.g., natural tags, photographs, DNA finger-prints), the probability of false links is often negligible, due to the caution in linkage procedures.

3.3.3 Bayesian approaches to record linkage

A Bayesian approach to record linkage is considered by [26], [35] and [36]. Specifically, in the approach of [26] and [39], the quantity of interest is a matrix-valued parameter C, which represents the true pattern of matches between the two lists. The sum of the elements of C is then an estimate of the number of true matches between the two lists, provided constraints on the parameter space of C which avoid multiple matches:

$$C_{ij} \in \{0,1\} \qquad \sum_{L_1} C_{ij} \leq 1 \qquad \sum_{L_2} C_{ij} \leq 1. \tag{3.9}$$

For a hierarchical Bayesian set-up, the top-level could be of the comparison vector γ, as in [26], where γ is assumed to have a multinomial distribution with parameters m or u, depending on C_{ij} being 1 or 0. The random vectors m and u are assumed independent from each other and independent of C and, for computational reasons, to follow a Dirichlet distribution. Alternatively, [39] and [59] propose models of the linking variables directly instead of the comparisons vector. A variation of the hit-miss model (see [13]) is used to model the process of possible distortion followed by uniform random re-alisation, i.e., of the observed values of the linking variables given their true values.

At the intermediate level, the number of true matches n_{11} is assumed to follow a hypergeometric distribution ([59]), given the target population size N and the sizes of the two lists, in the 2-list capture-recapture situation. The distribution of the matrix C is uniform over its possible outcome space, given n_{11}. Meanwhile, the vector of the true linking variables follows a suitable Urn distribution given n_{11}, generated by random sampling from the target population.

Finally, at the bottom level, the population distribution of the vector of true linking variables can be modelled by considering the target population as a random sample from an infinite super-population, with the associated parameters and their prior distribution. To complete the specification, one needs a prior distribution of N. Liseo and Tancredi [39] put $p(N) \propto 1/N^2$ truncated on $\{1, N^*\}$, where N^* is a reasonably large value. Tancredi and Liseo

[59] use $p_g(N) \propto \Gamma(N - g + 1)/N!$ for $g \geq 0$, where a larger g puts lower prior weight on the right tail.

Unlike the approach of Fellegi and Sunter [21], the Bayesian approach does not treat the matching of different pairs of records as unrelated to one another, which is logical. In practice, a linear constraint needs to be imposed after the initial linkage step ([33]), to avoid multiple links which is possible under the decision-rule approach of Fellegi and Sunter [21]. The Bayesian formulation more naturally takes into account the constraint via the matrix C. Finally, granted the Bayesian point of view to inference, the approach allows one to propagate the uncertainty of the linkage process to subsequent analysis of the linkage data, as will be illustrated in Section 3.4.

A practical difficulty with the Bayesian approach is the lack of scalability to large datasets, as in the case of census-like population-size estimation. Theoretically speaking, the choice of the prior distribution of the target parameter N seems somewhat arbitrary. In the motivating 2-list example of [59], we have $(n_{1+}, n_{+1}) = (34, 45)$, where $n_{11}^* = 25$ pairs have completely matched linking variables. The naïve DSE is 61, which is likely an overestimate due to the missing matches; whereas 45 would be a lower bound of N, assuming there is no erroneous enumeration. Being non-informative, the proposed prior does not accommodate such considerations. The resulting posterior median of N is 55, and the 97.5% quantile is 65, where the latter seems rather high compared to the naïve DSE. In any case, it is unclear how sensitive the results are towards the prior distribution, informative or non-informative.

Steorts et al. [55] propose an alternative Bayesian approach, which allows for linking and de-duplicating records from multiple lists simultaneously. The idea (similar to [44]) is to consider record linkage as a process of recognising the true latent "entities" underlying the separate files. Thus, each record refers to a target population unit (i.e., latent entity), which is represented by an integer from 1 to N_{max}, referred to as the linkage structure, where N_{max} is the total number of records in all the lists, so that in the extreme case the lists can have no common entities at all. All the records referring to the same entity are "linked". A uniform prior distribution is assumed for the linkage structure. The modelling approach is otherwise similar to that outlined above. A hybrid MCMC algorithm is used to generate the posterior distribution of the linkage structure. A rule of the most probable maximal matching set is introduced to guarantee transitivity in matching of multiple lists. See [57] for details.

For population-size estimation based on capture-recapture data, the approach of [57] in principle allows one to sample the set of cell counts, such as those in Tables 3.1 and 3.2, from their posterior distribution, given the observed linking variables. For instance, a latent entity attached to one record is only captured in that list, a latent entity attached to two records in different lists is captured in both these lists, and so on. Provided each record is attached to at least one latent entity, the set of cell counts can form the capture-recapture data for fitting the log-linear model, so as to derive the population-size estimate.

Tancredi et al. [60] extend [57] in the same Bayesian setting to allow for population-size estimation within a multiple-list record linkage and deduplication problem by considering the population size as a model parameter. They observed that under uniform prior on the units (i.e., independent random sampling with replacement from population N), the distribution of the sample labels given N induces a distribution on the partition space (i.e., the number of distinct latent individuals in the set) which depends on N, thus estimating the partition of the labels will permit at the same time to produce inference on N and estimate the linkage structure. The model is complete with the prior

$$P(N) = \frac{1}{\mathcal{Z}(g)N^g} \qquad N = 1,2,\ldots$$

where $\mathcal{Z}(g)$ is the Riemann zeta function. Again, the choice of prior (uniform) distribution of the linkage structure does not adapt to different scenarios of the capture-recapture probabilities. It might be more useful if one can set up and incorporate an informative prior distribution of the conditional cell probabilities of the log-linear model.

3.4 The DSE in the presence of linkage errors

3.4.1 The Ding and Fienberg estimator

The impact of linkage errors on capture-recapture estimates has been studied by [32]. In the presence of linkage errors, the coverage rates, e.g., (3.2) and (3.3), and the population-size estimate, e.g., (3.1), may be biased and need to be adjusted. In the 2-list case, a method to correct the Petersen estimator has been suggested by [19]. Let $n = n_{11}^* + n_{10}^* + n_{01}^*$ be given according to Table 3.1, but based on the linkage data, where $n_{1+}^* = n_{1+}$ and $n_{+1}^* = n_{+1}$. The face-value DSE using n_{11}^* instead of n_{11} is biased in general. The conditional likelihood of $(n_{11}^*, n_{10}^*, n_{01}^*)$ given n is

$$\mathcal{L}\left(\pi_{11}^*, \pi_{10}^*, \pi_{01}^* \mid n\right) = \frac{n!}{n_{11}^*! n_{10}^*! n_{01}^*!} \cdot \frac{\pi_{11}^{*\,n_{11}^*} \pi_{10}^{*\,n_{10}^*} \pi_{01}^{*\,n_{01}^*}}{(\pi_{11}^* + \pi_{10}^* + \pi_{01}^*)^n}. \tag{3.10}$$

where $\pi_{10}^* = \pi_1^* - \pi_{11}^* = \pi_1 - \pi_{11}^*$ and $\pi_{01}^* = \pi_2^* - \pi_{11}^* = \pi_2 - \pi_{11}^*$. To relate these parameters to the linkage errors, Ding and Fienberg (1994) make the following assumptions:

1. there is an assumed linkage direction, where L_1 is linked to L_2, i.e., for each record in L_1, one looks to find a link in L_2 if possible, but not the other way around;

2. true links between L_1 and L_2 are assigned with probability α;

3. false links involving matched records in L_1 are negligible, false links involving unmatched records in L_1 can occur with probability β.

Notice that the true match rate is the same as (3.8) except for linkage inter-dependence, but the false link rate is *not* the same as (3.7). Under these additional assumptions to the DSE, we have $\pi_{11}^* = \alpha\pi_1\pi_2 + \beta\pi_1(1 - \pi_2)$, where the two terms arise from n_{11} and n_{10}, respectively. Maximizing the conditional likelihood (3.10) with respect to π_1 and π_2, for given values of β and α, the estimated coverage of the first list is given by

$$\hat{\pi}_{1,DF} = \frac{-n_{11}^* + \beta(n_{11}^* + n_{10}^*)}{(\beta - \alpha)(n_{11}^* + n_{01}^*)} = \left(\frac{1}{n_{+1}}\right)\left(\frac{n_{11}^* - \beta n_{1+}}{\alpha - \beta}\right) \tag{3.11}$$

and that of the second list is given by

$$\hat{\pi}_{2,DF} = \frac{-n_{11}^* + \beta(n_{11}^* + n_{10}^*)}{(\beta - \alpha)(n_{11}^* + n_{10}^*)} = \left(\frac{1}{n_{1+}}\right)\left(\frac{n_{11}^* - \beta n_{1+}}{\alpha - \beta}\right) \tag{3.12}$$

Comparing (3.11) and (3.12) to (3.2) and (3.3), one recognizes the common term of (3.11) and (3.12) as the linkage-error adjusted estimate of the number of true matches. It follows that the conditional MLE, which is also the adjusted Peterson estimator, of N is given by

$$\tilde{N}_{DF} = (\alpha - \beta)\frac{n_{1+}n_{+1}}{n_{11}^* - \beta n_{1+}} = \frac{(\alpha - \beta)n_{11}^*}{n_{11}^* - \beta n_{1+}}\tilde{N}_P^* \tag{3.13}$$

where $\tilde{N}_P^*$ is the face-value DSE using n_{11}^* directly; see [19].

3.4.2 The modified Ding and Fienberg estimator

Di Consiglio and Tuoto [16] propose a generalization, which relaxes the restriction of one-way linkage. Linking of administrative sources motivates this need. Notice that Ding and Fienberg [19] address the traditional census under-coverage evaluation, where the linkage between the census enumeration and the post-enumeration survey worked in one direction. When one-way linkage is not the case, false links can occur for unmatched records in L_1, as well as those in L_2, so that we have instead $\pi_{11}^* = \alpha\pi_1\pi_2 + \beta\pi_1(1 - \pi_2) + \beta\pi_2(1 - \pi_1)$, with an additional term due to the latter. Again, this false link rate differs to (3.8); neither is it the same as that of [19], since the two have different denominators. De Wolf et al. (2018) allow the probabilities of missing matches to be different under two-way linkage.

Retaining the other assumptions about the linkage errors as above, maximizing the conditional likelihood (3.10), for given values of β and α, the Modified Ding and Fienberg (MDF) estimators of the coverage rates are given by, respectively,

$$\hat{\pi}_{1,MDF} = \frac{2\beta n_{11}^* + \beta x_{10}^* + \beta n_{01}^* - n_{11}^*}{(2\beta - \alpha)(n_{11}^* + n_{01}^*)} = \left(\frac{1}{n_{+1}}\right)\left(\frac{n_{11}^* - \beta(n_{1+} + n_{+1})}{\alpha - 2\beta}\right) \tag{3.14}$$

$$\hat{\pi}_{2,MDF} = \frac{2\beta n_{11}^* + \beta n_{10}^* + \beta n_{01}^* - n_{11}^*}{(2\beta - \alpha)(n_{11}^* + n_{10}^*)} = \left(\frac{1}{n_{1+}}\right)\left(\frac{n_{11}^* - \beta(n_{1+} + n_{+1})}{\alpha - 2\beta}\right) \qquad (3.15)$$

Again, the common term of (3.14) and (3.15) is the linkage-error adjusted estimate of n_{11}, so that the MLE, which is also the adjusted Peterson estimator, of N is given by

$$\tilde{N}_{MDF} = (\alpha - 2\beta)\frac{n_{1+}n_{+1}}{n_{11}^* - \beta(n_{1+} + n_{+1})} = \frac{(\alpha - 2\beta)n_{11}^*}{n_{11}^* - \beta(n_{1+} + n_{+1})}\tilde{N}_P^* \qquad (3.16)$$

As presented above, the DF and MDF estimators are both based on the assumption that linkage errors are constant overall. If this assumption holds at least in sub-groups, the estimators can be applied within strata in which the linkage error probabilities (and capture probabilities) are more homogeneous than in the whole population.

When the error rates are assumed to be known, one may consider the true values n_{11}, n_{10}, n_{01} to be obtained deterministically by algebra from $n_{11}^*, n_{10}^*, n_{01}^*$. The variance estimator for $\tilde{N}_{MDF}$ is then the same as the standard DSE variance estimator (see [63]), which is given by

$$\hat{V}(\tilde{N}_{MDF}) = N\frac{(1 - \hat{\pi}_1)(1 - \hat{\pi}_2)}{\hat{\pi}_1\hat{\pi}_2} \qquad (3.17)$$

where $\hat{\pi}_1 = \hat{n}_{11}/n_{+1}$ and $\hat{\pi}_2 = \hat{n}_{11}/n_{1+}$. The same approach is taken in [58] for correcting the capture-recapture two-sample abundance estimator to account for false negative errors in identification. Furthermore, they introduce a bootstrap approach for estimating the variance of the corrected estimator. The bootstrap allows incorporating the estimation uncertainty of the linkage error rates.

3.4.3 Some remarks

Below we provide some remarks on the DSE in the presence of linkage errors, and make a comparison of its bias and variance to the DSE in the absence of linkage errors.

Firstly, under the standard motivation for the DSE estimator, *both* lists are considered to be random, independently enumerated, each with a constant coverage rate, denoted by τ_1 and τ_2, respectively. We have then

$$\begin{cases} E(n_{1+}) = N\tau_1 \\ E(n_{+1}) = N\tau_2 \\ E(n_{11}) = N\tau_1\tau_2 \end{cases} \Rightarrow \frac{E(n_{1+})E(n_{+1})}{E(n_{11})} = N \Rightarrow \hat{N} = \frac{n_{1+}n_{+1}}{n_{11}}$$

As demonstrated in [65] and [66], a simpler motivation, which requires fewer assumptions and is more suitable to lists arising from administrative sources, is to treat one of the lists, say List 1 as *fixed*. Assume random n_{+1}, with constant capture rate throughout the target population, we have, conditional on n_{1+},

$$\left\{ \begin{array}{l} E(n_{1+}|n_{1+})=n_{1+} \\ E(n_{+1}|n_{1+})=E(n_{+1})=N\tau \\ E(n_{11}|n_{1+})=n_{1+}\tau \end{array} \right. \Rightarrow \frac{n_{1+}E(n_{+1})}{E(n_{11}|n_{1+})} = N \Rightarrow \hat{N} = \frac{n_{1+}n_{+1}}{n_{11}}$$

We shall adopt this simplifying conditional perspective in the following.

Let α be the probability that a match between Lists 1 and 2 is included in the linked joint subset. Let β be the probability that an unmatched record in List 1 or 2 is included in the linked joint subset. We have

$$E(n_{11}^*|n_{1+}) = \alpha E(n_{11}|n_{1+}) + [n_{1+} - E(n_{11}|n_{1+})]\beta + [E(n_{+1}) - E(n_{11}|n_{1+})]\beta$$
$$= (\alpha - 2\beta)E(n_{11}|n_{1+}) + [n_{1+} + E(n_{+1})]\beta$$

Given (α,β), we can obtain the following estimators

$$\hat{n}_{11} = \hat{n}_{11}(\alpha,\beta) = \frac{n_{11}^* - \beta(n_{1+} + n_{+1})}{(\alpha - 2\beta)}$$

$$\hat{N}(\alpha,\beta) = \frac{n_{1+}n_{+1}}{\hat{n}_{11}(\alpha,\beta)}$$

which is identical to the MDF estimator in (3.16). The special case of $\beta = 0$ is of particular interest, for instance, in animal abundance estimation, under which we have

$$\hat{N}(\alpha) = \hat{N}(\alpha,0) = \frac{\alpha n_{1+}n_{+1}}{n_{11}^*}$$

Below we make an explicit comparison between the adjusted DSE $\hat{N}(\alpha)$ and the standard DSE *conditional on* n_{+1} in addition to n_{1+}. This is the usual case in which the DSE is applied, provided the counts in each single source are observed and considered given. The difference between with and without linkage errors is between observing n_{11}^* or n_{11}. We have, for the standard DSE $\hat{N}$,

$$\hat{N} = \frac{n_{1+}n_{+1}}{n_{11}} \approx \frac{n_{1+}n_{+1}}{E(n_{11}|n_{1+},n_{+1})} - \frac{n_{1+}n_{+1}}{E(n_{11}|n_{1+},n_{+1})^2}[n_{11} - E(n_{11}|n_{1+},n_{+1})]$$
$$+ \frac{1}{2}\frac{2n_{1+}n_{+1}}{E(n_{11}|n_{1+},n_{+1})^3}[n_{11} - E(n_{11}|n_{1+},n_{+1})]^2$$

$$E(\hat{N}|n_{1+},n_{+1}) \approx \frac{n_{1+}n_{+1}}{E(n_{11}|n_{1+},n_{+1})} - \frac{n_{1+}n_{+1}}{E(n_{11}|n_{1+},n_{+1})^3}V(n_{11}|n_{1+},n_{+1})$$

$$V(\hat{N}|n_{1+},n_{+1}) \approx \frac{(n_{1+}n_{+1})^2}{E(n_{11}|n_{1+},n_{+1})^4}V(n_{11}|n_{1+},n_{+1})$$

We have $n_{11} \sim Bin(n_{+1}, n_{1+}/N)$ conditional on (n_{+1}, n_{1+}), because the probability is the same for a record in or out of List 1 to be enumerated in List 2. Therefore, $E(n_{11}|n_{1+}, n_{+1}) = n_{1+}n_{+1}/N$ and $V(n_{11}|n_{1+}, n_{+1}) = n_{+1}(\frac{n_{1+}}{N})(1 - \frac{n_{1+}}{N})$, such that

$$E(\hat{N}|n_{1+}, n_{+1}) \approx N + \frac{N}{\hat{N}^2}n_{+1}\frac{n_{1+}}{N}\left(1 - \frac{n_{1+}}{N}\right) = N + O(1)$$

$$V(\hat{N}|n_{1+}, n_{+1}) \approx \frac{N^2}{\hat{N}^2}n_{+1}\frac{n_{1+}}{N}\left(1 - \frac{n_{1+}}{N}\right)$$

where we assume $\frac{n_{1+}}{N} = \frac{n_{+1}}{N} = \frac{\hat{N}}{N} = O(1)$ as $N \to \infty$. Next, similarly for $\hat{N}(\alpha)$, we have

$$E(\hat{N}(\alpha)|n_{1+}, n_{+1}) \approx \frac{\alpha n_{1+}n_{+1}}{E(n_{11}^*|n_{1+}, n_{+1})} - \frac{\alpha n_{1+}n_{+1}}{E(n_{11}^*|n_{1+}, n_{+1})^3}V(n_{11}^*|n_{1+}, n_{+1})$$

$$V(\hat{N}(\alpha)|n_{1+}, n_{+1}) \approx \frac{(\alpha n_{1+}n_{+1})^2}{E(n_{11}^*|n_{1+}, n_{+1})^4}V(n_{11}^*|n_{1+}, n_{+1})$$

Since $n_{11}^* \sim Bin(n_{11}, \alpha)$ conditional on n_{11} and $n_{11} \sim Bin(n_{+1}, n_{1+}/N)$ conditional on (n_{+1}, n_{1+}), we have $n_{11}^* \sim Bin(n_{+1}, \alpha n_{1+}/N)$ conditional on (n_{+1}, n_{1+}). Therefore, $E(n_{11}^*|n_{1+}, n_{+1}) = \alpha n_{1+}n_{+1}/N$ and $V(n_{11}^*|n_{1+}, n_{+1}) = n_{+1}(\alpha n_{1+}/N)(1 - \alpha n_{1+}/N)$, such that

$$E(\hat{N}(\alpha)|n_{1+}, n_{+1}) \approx N + \frac{N}{\hat{N}(\alpha)^2}n_{+1}\frac{\alpha n_{1+}}{N}\left(1 - \frac{\alpha n_{1+}}{N}\right) = N + O(1)$$

$$V(\hat{N}(\alpha)|n_{1+}, n_{+1}) \approx \frac{N^2}{\hat{N}(\alpha)^2}n_{+1}\frac{\alpha n_{1+}}{N}\left(1 - \frac{\alpha n_{1+}}{N}\right)$$

It follows that both the conditional bias and variance of the $\hat{N}(\alpha)$, i.e., in the presence of linkage errors, are of the same order as those of $\hat{N}$, i.e., in the absence of the linkage errors. However, provided both n_{1+}/N and α are close to 1, we have

$$V(n_{11}^*|n_{1+}, n_{+1}) > V(n_{11}|n_{1+}, n_{+1})$$

because $\alpha n_{1+}/N$ is then closer to 0.5 than n_{1+}/N. In other words, one can expect some loss of efficiency due to the presence of linkage errors.

We now return to the general case of $\widehat{N}(\alpha, \beta)$. To start with, observe that

$$\hat{n}_{11} = n_{11}^* + (1-\alpha)\hat{n}_{11} - (n_{1+} + n_{+1} - 2\hat{n}_{11})\beta$$

That is, we have $\hat{n}_{11} = n_{11}^*$ when $(1-\alpha)\hat{n}_{11} = (n_{1+} + n_{+1} - 2\hat{n}_{11})\beta$, i.e., if the two types of linkage errors happen to 'cancel each other'. To gauge the expectation and variance of $\hat{n}_{11}$, let

$$n_{11}^* = n_{11}^*(1,1) + n_{11}^*(0,1) + n_{11}^*(1,0)$$

which arise, successively, from the matches between Lists 1 and 2, the unmatched records in List 2 but not 1, and those in List 1 but not 2. Conditional on (n_{+1}, n_{1+}), we have

$$n_{11}^*(1,1) \sim Bin(n_{+1}, \alpha n_{1+}/N)$$

$$n_{11}^*(0,1) \sim Bin(n_{+1}, \beta(1 - n_{1+}/N)),$$

and similarly

$$n_{11}^*(1,0) \sim Bin(n_{1+}, (1 - n_{+1}/N)\beta).$$

It follows that

$$E(n_{11}|n_{1+}, n_{+1}) = (\alpha - 2\beta)\frac{n_{1+}n_{+1}}{N} + (n_{1+} + n_{+1})\beta$$

$$E(\hat{n}_{11}|n_{1+}, n_{+1}) = \frac{n_{1+}n_{+1}}{N}$$

$$E\left(\hat{N}(\alpha, \beta)|n_{1+}, n_{+1}\right) = N + \frac{N}{\hat{N}(\alpha, \beta)^2}V(\hat{n}_{11}|n_{1+}, n_{+1})$$

Observe that $n_{11}^*(1,1)$, $n_{11}^*(0,1)$ and $n_{11}^*(1,0)$ are not independent of each other, since for example, smaller n_{11} by chance makes smaller $n_{11}^*(1,1)$ and bigger $n_{11}^*(0,1)$ and $n_{11}^*(1,0)$ more probable at the same time.

While $V(\hat{n}_{11}|n_{1+}, n_{+1})$ seem intractable analytically, it is possible to calculate it by Monte Carlo simulations. We conjecture that $V(\hat{n}_{11} \mid n_{1+}, n_{+1})/N = O(1)$ asymptotically as $N \to \infty$, in which case the bias and variance of the $\hat{N}(\alpha, \beta)$ would be of the same order as those of the standard DSE $\hat{N}$.

3.4.4   Examples

Below we present some 2-list examples in the presence of linkage errors.

Example 1. Consider the data in Tables 3.3 and 3.4 from [23], in three sources: the 1990 U.S. census, the corresponding post enumeration survey (PES), the administrative list supplement (ALS), for the PES sampling strata 11 in St. Louis. The Census-PES capture-recapture data are given in Table 3.3, whereas results of the Matching Error Study ([42]) are shown in Table 3.4.

TABLE 3.3
Census and PES sample data for stratum 11, St. Louis, 1990 U.S. Census

		Census	
		Present	*Absent*
PES	*Present*	487	129
	Absent	217	-

TABLE 3.4
St. Louis Rematch Study

Original match classification	Rematch Classification			Total
	Matched	Not matched	Unresolved	
Matched	2667	7	8	2682
Not matched	9	427	30	466
Unresolved	0	7	20	27
Total	2676	441	58	3175

Treating the rematch results as the truth and ignoring the unresolved cases, as in [23], the missing match rate is $1 - \widehat{\alpha} = 9/(2667+9) = 0.3363\%$, and the false link rate is $\widehat{\beta} = 7/(7+427) = 1.6129\%$. The standard DSE resulting from Table 3.3 is 890, with a standard error (SE) 10.25. Incorporating the linkage errors, we obtain $\tilde{N}_{DF} = 891$ (with SE=10.29). Due to the low error rates the adjustment is small, and the two estimates do not differ significantly in light of the SEs. Notice that the naïve DSE is actually adjusted upwards due to the very low rate of missing matches.

Plugging in the two error rates directly in (3.16), the MDF estimate is $\tilde{N}_{MDF} = 898$ (with SE=10.64). However, given that Table 3.3 is actually generated by one-way linkage of PES to Census, the MDF estimate is not valid in this case, as it would suggest that there are $5.5 \approx (129 + 217) \cdot \hat{\beta}$ false links among the 487 actual links made, whereas the number is $2 \approx 129 \cdot \hat{\beta}$ according to the one-way linkage procedure. The MDF estimate and its SE are only calculated here to illustrate the computation involved.

Example 2. Consider the case described in [16], with data from two registers of death incidents caused by road accidents. A complete analysis of the linkage result by clerical review is possible due to their small sizes. The Road Accident Register (RAR), recording dynamics and circumstances of the road accidents, is linked with the Registry on Causes of Death (RCD). The linkage procedure is not straightforward: a common personal identifying code is not available and, moreover, since RAR reference units are the road accidents, personal identifying variables (i.e., names, surnames, ages) are sometimes missing or mistaken when more than one person is involved. The data from RAR consists of 4237 records, and that from RCD has 4642 records. The variables used for the linkage include the road traffic victim/dead person's name, surname and age, and the accident/death day, month, municipality and province. The data sizes do not require reduction procedures and all the possible pairs of records are examined. The entire space of possible links is also explored by clerical review for links missed by the probabilistic procedure. The linkage procedure identifies 3129 linked records. The estimated Fellegi-Sunter (FS) linkage error rates by (3.7) and (3.8) are given by $\hat{\beta}=0.00$ and $1 - \hat{\alpha}=0.15$. The post-linkage clerical review of the linkage status allows us to evaluate the true linkage error rates (Table 3.5), according to which the true $1 - \alpha$ is 0.1141 and β is 0.0011. Moreover, using the true links after

TABLE 3.5

Comparison between true linkage status and probabilistic linkage results

		True Linkage Status		Total
		Match	Non-Match	
Probabilistic	Link	3127	2	3129
Linkage	Non-Link	403	-	2621
Total		3530	1819	

clerical review, the DSE of the total number of road deaths is 5572, whereas the naïve DSE is 6286.

The population-size estimate obtained by the MDF estimator is 5571 if true linkage error rates are used, whereas it is 5330 using the FS estimates. As can be expected, the DF estimate is virtually the same as the MDF estimate, now that β is almost zero. However, the DF estimator would have been invalid given a non-zero false link rate, because the linkage procedure is not one-way in this case. The naïve DSE is an over-estimate, almost entirely due to the missed matches, which is the dominating source of bias.

It is worthwhile to notice that the evaluation of β and $1-\alpha$ is not straightforward, when the true linkage status is only known for a training set of links but not the entire space of all possible links. For instance, the well-known method by [5] only provides estimate of β. The method proposed by [61] estimates both β and $1-\alpha$ without strong distributional assumptions on the linkage weights.

Example 3. This example illustrates Bayesian uncertainty propagation for counting wild animal population. The data is reported in [38], collected from a study of marbled salamanders, in the Shenandoah Valley of Virginia, USA (details in [2]). In this study, the captures are realized by means of a photograph of the dorsal pattern, and computer-assisted pattern matching software is used to identify individuals and to construct the capture-recapture data (Table 3.6). Two captures are realized before and after migration to and from a pond where adult salamanders lay eggs in early September.

TABLE 3.6

Capture-recapture of marbled salamanders, in the Shenandoah Valley of Virginia, USA

	Second Capture		Total
First Capture	Present	Absent	
Present	174	22	196
Absent	363	-	
Total	537		

As mentioned before, in ecological studies like this one, the main risk of errors is the misidentification of natural tags, i.e., the rate of missing matches

α, which generates "ghost" records according to the terminology of Link et al. [38]. The identified matches will then be too low, and the population-size estimate too high.

There are four unknown parameters for the data in Table 3.6, namely the capture probabilities on the two occasions, π_1 and π_2 respectively, the population size N and the misidentification error rate α. The model is over-parameterized. Link et al. (2010, p. 183) considered several options:

1. to ignore α and apply the naïve DSE;

2. to fit a model with α under the restriction of $\pi_1=\pi_2$;

3. to conduct a Bayesian analysis with a solicited prior on α.

The second option is clearly not defendable given the data in Table 3.6. The first option is actually a special case of the third option, with a point-mass prior on α. After consulting the field experts, [38] settled on uniform priors for π_1, π_2 and N, and the Beta(19,1) prior distribution for α suggesting a high but less than perfect identification rate. However, Uniform, Beta(10,1), Beta(19,1) priors for α lead to nearly identical 95% credible intervals, see [38] for details.

Table 3.7 provides a summary of the results. The posterior distribution of N is obtained by the Gibbs sampler ([38]). The median, the 2.5th and the 97.5th percentiles of the posterior distribution of N are given for different priors for α. For comparison, also the MLE and the associated confidence intervals are calculated, treating the expected α under its prior distribution as if it were known. Notice that the MLE in the last row corresponds to the naïve DSE, which ignores the identification errors, whereas the adjusted MLE cannot be computed for expected value of α in the first row, because the adjusted number of true matches is larger than the total from the first capture.

TABLE 3.7
Estimates for N under alternative priors for α

Prior for α	Posterior median (2.5th, 97.5th percentiles) of N	Expected value of α	MLE and 0.05% confidence interval
Beta(1,1)	569 (538, 616)	0.50	-
Beta(10,1)	573 (538, 618)	0.91	550 (540, 560)
Beta(19,1)	578 (539, 620)	0.95	575 (557, 592)
Beta(100,1)	595 (557, 628)	0.99	599 (575, 623)
Beta(∞,1)	606 (585, 636)	1	605 (580, 630)

We notice that the length of the confidence interval decreases with α, which is understandable if α were known, because it then means that the number of true matches is known and increasing while the two list sizes are fixed. However, treating the stipulated α as if it were known is clearly misleading in this case. The Bayesian posterior uncertainty increases as the prior expectation of α decreases, due to the fact that the prior distribution is becoming less concentrated around its center. The results for less certain priors than Beta(19,1) suggest that (i) the posterior 95% interval is robust against the choice of the prior, and (ii) the posterior median is a more robust choice for point estimation than the posterior mean.

Example 4. A simulation study is proposed in [16] in order to compare the estimators in different linkage scenarios, highlighting the bias of the standard capture-recapture estimates. The fictitious data used in the simulation mimic the register under-coverage and the presence of errors in the linking variables [41]. The two lists were randomly generated according to the following coverage probabilities, $\tau_1 = 0.930$ and $\tau_2 = 0.924$, respectively. The two lists were linked assuming three different scenarios. The Gold scenario uses linking variables with the highest identifying power, namely, *Name, Surname, Complete Date of Birth*. This gives the best results in terms of the error rates, where $\alpha=0.939$ and $\beta=0.001$ on average over the replications. The Silver scenario represents the situation where the strongest identifying variables – namely, *Name* and *Surname* – are not available, as for example, may be the case due to privacy protection. The linkage procedure is based on the *Complete Date of Birth*. This causes higher linkage error rates than in the Gold scenario, where $\alpha=0.851$ and $\beta=0.101$ on average. Finally, the Bronze scenario is the most unfavorable in terms of linkage errors, due to additional typos and missing values of the linking variables. The resulting average values of α and β were 0.833 and 0.108, respectively. The naïve DSE, DF and MDF estimators were compared to each other, where the DF and the MDF estimators were computed using the true values of α and β in each replication. The results were also compared to the true DSE unaffected by linkage errors.

In the Gold scenario, where the false link error is nearly absent, the adjusted estimators reduce the bias of the naïve DSE and, as can be expected, the DF and the MDF are very close in this situation. In the Silver scenario, where the false link rate β is no longer negligible, the MDF clearly outperforms the alternative estimators, where DF estimator is inappropriate since the linkage is not one-way. The improvement by the MDF estimator is even more evident in the Bronze scenario with a higher false link rate. See [16] for more details on the replication settings and the results.

TABLE 3.8
Table for 3-list linkage cell counts

	List 1			
	Present		Absent	
	List 3		List 3	
List 2	Present	Absent	Present	Absent
Present	n^*_{111}	n^*_{110}	n^*_{011}	n^*_{010}
Absent	n^*_{101}	n^*_{100}	n^*_{001}	n^*_{000}

3.5 Linkage-error adjustments in the case of multiple lists

This section investigates the effect of linkage errors and proposes adjustments of the population-size estimators based on multiple-recapture log-linear models introduced in Section 3.2.

3.5.1 Log-linear model-based estimators

We start with the 3-list linkage data in Table 3.8, where n^*_{ijk} is the cell count based on linkage data and π^*_{ijk} the corresponding cell probability, for $i,j,k=1,0$, and n^*_{000} is unobservable by definition. The table is arranged in the same way as the true capture-recapture data in Table 3.2. Linkage errors "perturb" the data in Table 3.2, so that n^*_{ijk} may differ to n_{ijk}, except for the list totals, n_{1++}, n_{+1+} and n_{++1}. Let n be the number of all observed distinct units, which is also assumed to be unaffected by linkage errors.

Fienberg and Ding [23] propose a correction of the log-linear model that considers the possible transitions from the true configuration n to the observed one taking into account only the missing links. They assume that:

(i) there are no erroneous matches in the linkage process;

(ii) a transition can only go downwards by at most one level;

(iii) the probability of remaining at the original state (no missing error) equals α and the probability of a transition to any of the other possible states is equal to $(1 - \alpha)/(m - 1)$, where m is the number of all possible states to which transitions are possible and allowed.

For example, an individual truly recorded in all the three lists (111) can produce the following patterns {(110), (001)} or {(101), (010)} or {(011), (100)} with equal probability $(1 - \alpha)/3$.

Similarly, {(110), (001)} can either be observed as {(110), (001)} or {(100), (010),(001)}, and so on for the other true match patterns. This formulation can be easily extended to the general K-list data. The permissible transition structure gives rise to an equation that relates the linkage-table cell probabilities to those of the true table, denoted by $\pi^* = M\pi$, and multiplying by N, one obtains the expected cell counts as $E(n^*|n) = Mn$.

To estimate the population size N, Fienberg and Ding [23] suggest to compute the MLE of π from the conditional likelihood based on n^* given n, as in Section 3.4.1. As shown by [51], the conditional and unconditional MLEs are both consistent, under suitable regularity conditions. The MLEs of n_{000} and N can then be obtained as explained in [22], as mentioned in Section 3.2.

Di Consiglio and Tuoto [17] propose an extension to allow for false links in additional to missing matches. Moreover, they apply an error model taking into account the proposed operational model for the 3-list linkage procedures, as described in Section 3.3.1, where Lists 1 and 2 are linked at first, and then with List 3 afterwards.

Accordingly, let $1 - \alpha_1$ be the probability of missing a match in the first linkage and assuming that the probability of missing a match in the second linkage is independent to the results of the first linkage operation, $1 - \alpha_2$. Similarly let β_1 and β_2 be the false link probabilities of the two linkages, respectively, assuming that the probability of an incorrect link does not depend on the result of the first operation. Differentiating the errors in the two operations of linkage is motivated by varying key variables quality in the different lists.

Take again n_{111}. When only missing matches are allowed, the possible transitions from the true match pattern (111) to the observed ones are:

{(111)}	with probability	$\alpha_1\alpha_2$
{(110), (001)}	with probability	$\alpha_1(1-\alpha_2)$
{(101), (010)} or {(100), (011)}	with probability	$(1-\alpha_1)\alpha_2/2$
{(100), (010), (001)}	with probability	$(1-\alpha_1)(1-\alpha_2)$

Putting together all the other true match patterns similarly, Table 3.9 gives the transition matrix from the true data in Table 3.2 to the observed data in Table 3.8.

TABLE 3.9
Transition matrix from real to observed pattern with only missing links

	(111)	(110)	(101)	(100)	(011)	(010)	(001)
(111)*	$\alpha_1\alpha_2$						
(110)*	$\alpha_1(1-\alpha_2)$	α_1					
(101)*	$(1-\alpha_1)\alpha_2/2$		α_2				
(100)*	$(1-\alpha_1)(2-\alpha_2)/2$	$1-\alpha_1$	$1-\alpha_2$	1			
(011)*	$(1-\alpha_1)\alpha_2/2$				α_2		
(010)*	$(1-\alpha_1)(2-\alpha_2)/2$	$1-\alpha_1$			$1-\alpha_2$	1	
(001)*	$1-\alpha_2$		$1-\alpha_2$		$1-\alpha_2$		1

Next, the transition matrix can be extended to include the false link errors (Table 3.10). Moreover, at each phase, one may suppose that whenever a true match is missed, a false link cannot occur, because this event occurs when at least two errors are made simultaneously, similarly to the 2-list situation, Sections 3.4.1 and 3.4.2. Finally, as mentioned already in Section 3.3.2, note that in Table 3.10, the false link rates are evaluated using the denominator $(N_1 - n_{11})+(N_2 - n_{11})$, i.e., the number of unmatched record in either List 1 or 2 and $(N_{11} - n_{111})+(N_3 - n_{111})$, where N_{11} denote the new list after the first

TABLE 3.10

Transition matrix from real to observed pattern with missing and false links

	(111)	(110)	(101)	(100)	(011)	(010)	(001)
(111)*	$\alpha_1\alpha_2$	$\alpha_1\beta_2$	$\alpha_2\beta_1$	$\beta_1\beta_2$	$\alpha_2\beta_1$		
(110)*	$\alpha_1(1-\alpha_2)$	$(\alpha_1)(1-\beta_2)$	$\beta_1(1-\alpha_2)$	$(\beta_1)(1-\beta_2)$	$\beta_1(1-\alpha_2)$		
(101)*	$\frac{(1-\alpha_1)(\alpha_2)}{2}$	$(1-\alpha_1)\beta_2/2$	$(1-\beta_1)(\alpha_2)$	$(1-\beta_1)(\beta_2)$			
(100)*	$\frac{(1-\alpha_1)(2-\alpha_2)}{2}$	$(1-\alpha_1)\left(1-\frac{\beta_2}{2}\right)$	$(1-\beta_1)(1-\alpha_2)$	$(1-\beta_1)(1-\beta_2)$	$-\beta_1$		
(011)*	$\frac{(1-\alpha_1)(\alpha_2)}{2}$	$(1-\alpha_1)\beta_2/2$		$-\beta_1$	$(1-\beta_1)\alpha_2$		
(010)*	$\frac{(1-\alpha_1)(2-\alpha_2)}{2}$	$(1-\alpha_1)\left(1-\frac{\beta_2}{2}\right)$	$-\beta_1$	$-\beta_1$	$(1-\beta_1)(1-\alpha_2)$	1	
(001)*	$1-\alpha_2$	$-\beta_2$	$1-\alpha_2$	$-\beta_2$	$1-\alpha_2$		1

linkage operation, respectively in the two steps. Whereas the missed matches probabilities are evaluated using the denominator $(N_1 - n_{11}^*)+(N_2 - n_{11}^*)$, and $(N_{11} - n_{111}^*)+(N_{32} - n_{111}^*)$.

TABLE 3.11
Transition matrix from real to observed pattern for two lists

	11	10	01
11*	α_1	β_1	β_1
10*	$1 - \alpha_1$	$1 - \beta_1$	$-\beta_1$
01*	$1 - \alpha_1$	$-\beta_1$	$1 - \beta_1$

Having specified the transition matrix, one may follow the same estimation approach as Fienberg and Ding (1996) outlined above. Notice that the 2-list situation in Section 3.4.2 can now be represented by the transition matrix in Table 3.11. The adjustment approaches in Section 3.4 are thus special cases of the approach described here.

3.5.2 An alternative modelling approach

Chipperfield et al. [12] propose an EM-algorithm based approach to the analysis of contingency tables arising from record linkage in the following setting. Let X be a categorical variable available in File 1, and Y that in File 2. Fix the arrangement of records in File 1, so that we observe X_a for each record a in File 1. Consider a linked pair $(a,b) \in M^*$, where $X_a = x$ and $Y_a^* = c$, and c is the value associated with the record (from File 2) which is linked to the record a, denoted by $w_{c|x}^* = 1$, and let $w_{c|x}^* = 0$ for any other value $y \neq c$. Let $w_{c|x} = 1$ if the true matched record to a has $Y = c$, and $w_{c|x} = 0$ otherwise. The idea is now to model the conditional probability of $w_{c|x}$ given $w_{c|x}^*$ and $X = x$. Chipperfield et al. adopt the following model

$$\Pr(w_{c|x} = 1 | w_{c|x}^* = 1, X = x) = 1 - \beta_{xc} \quad \Pr(w_{c|x} = 1 | w_{c|x}^* = 0, X = x) = \beta_{xc}\pi_{c|x}$$

where β_{xc} is the probability that a link with observed values (x,c) is a false match, and $\pi_{c|x}$ is the true marginal conditional probability of $Y = c$ given $X = x$. Notice that the assumption is essentially the same as the hit-miss model of [13]. Notice also that the false match probability β_{xc} is similar to that of [8] for regression analysis, but has a *different* definition to any of the false link rates discussed so far. In any case, under this assumed model, one can now calculate the conditional expectation $E(\mathbf{n}|\mathbf{n}^*)$, where $\mathbf{n}^*$ is the set of observed cell counts based on the linkage data, and $\mathbf{n}$ that of the unobserved true data. This yields the E-step of the EM-algorithm. At the M-step, one then maximizes the postulated model of (X, Y) using $E(\mathbf{n}|\mathbf{n}^*)$ as the data. Iteration till convergence is necessary.

It follows from the definition of $w_{c|x}^*$ that one only looks at linked pairs of records. Therefore, a key premise of this approach is that there are incorrect links but there are no unlinked records. This condition is clearly violated in the case of capture-recapture data, where by definition there are *always* unlinked records. To give an appreciation of the problem, let us consider how one might have attempted the same modelling approach in the 2-list situation. Let $x = 1$ if a record in List 1 has a true match in List 2, and $x = 0$ otherwise. Let $y = 1$ and 0 denote that for a record in List 2. Similarly, let $x^* = 0$ if a record in List 1 is linked to a record in List 2, and $x^* = 0$ otherwise. Likewise for y^*.

Take an unlinked record a in List 1, i.e., $x^* = 0$. Under the present set-up, it contributes to n_{10}^* in $\mathbf{n}^* = (n_{11}^*, n_{10}^*, n_{01}^*)$. If a is actually an unmatched record, then $x = 0$ in truth, otherwise $x = 1$. Notice that it is *not* possible to condition on the true x here. Next, in case $x = 1$, the matched record, say, b from List 2 can either have $y^* = 0$ if it remains unlinked, or $y^* = 1$ if it is mistakenly linked to another record from List 1. One may take the probability of the second event to be negligible as it involves multiple linkage errors, similar to the view taken by Ding and Fienberg (1994). Thus, given $x^* = 0$, suppose one stipulates two possible events: (i) $x = 0$ for a, which contributes to n_{10} in $\mathbf{n}$; (ii) $x = 1$ for a, $y = 1$ and $y^* = 0$ for some b in List 2, such that the two records contribute to n_{11}. In particular, the probability of event (ii) can be given by

$$P\big[(ab) \in M | (ab) \in U^*\big] = P\big[(ab) \in U^* | (ab) \in M\big] \frac{\Pr[(ab) \in M]}{\Pr[(ab) \in U^*]} = \frac{(1-\alpha)n_{11}}{n_1 n_2 - n_{11}^*}$$

with α by (3.3), and both $\Pr[(ab) \in M]$ and $\Pr[(ab) \in U^*]$ have the denominator $n_1 n_2$. The probability of event (i) is $1 - P\big[(ab) \in M | (ab) \in U^*\big]$.

Similarly for an unlinked record b in List 2, i.e., given $y^* = 0$, one may stipulate only two events: (i') $y = 0$ for b; (ii') $y = 1$ for b, $x = 1$ and $x^* = 0$ for some a in List 1. Yet it would not be correct to write down the same probability of event (ii') as above, because the record b may or may not have already been accounted for in event (ii), for some unlinked record a in List 1. Suppose one first goes through the shorter unlinked list, and calculates the contribution to $\mathbf{n}$ given the observed contribution to $\mathbf{n}^*$, according to the probabilities stipulated above. Having done so, one runs into the following dilemma. On the one hand, there are still unlinked records yet to be accounted for; on the other hand, one has 'used up' all the records that can account for them. It seems that there are some difficulties to apply the approach of [12] in situations where there are unlinked records.

3.5.3 A Bayesian proposal

Assuming only missing matches but no false links, Link et al. [38] generalize the 2-list approach (Section 3.4) to K-list situations. The capture-identification history of each individual can be represented by a vector, where the component corresponding to a given capture occasion can take values over the following possibilities:

(i) presence on the given occasion and correctly identified as such;

(ii) presence on the given occasion but unidentified as such;

(iii) absence on the given occasion.

The true capture-identification history of all the individuals gives rise to a set of unobserved true capture-recapture counts, denoted by n, as well as the observed linkage-data counts of the same structure, denoted by n^*. Under the assumption of only missing matches but no false links, we obtain the same linear relationship $E(n^*|n) = Mn$ as in Section 3.5.1. Link et al. [38] suggest Gibbs sampling under the Bayesian framework. In any case, the proposal has essentially the same modelling requirement, in terms of the specification of transition matrix M, as under the approach in Section 3.5.1. We have already discussed Bayesian method together with the linkage approach in Section 3.3.3. In fact record linkage and estimation are not treated as a separate process and full hierarchical Bayesian approach would require specification of the complete model addressing record linkage and parameter estimation.

3.5.4 Examples

Example 5. Consider Example 1 again, with an administrative list supplement (ALS) in addition to the Census and PES enumerations ([23]). The 3-list capture-recapture linkage data are given in Table 3.12.

TABLE 3.12
Three-list data for stratum 11, St. Louis, 1990 U.S. Census

	Census (List 1)			
	Present		Absent	
	PES (List 2)		PES (List 2)	
ALS (List 3)	Present	Absent	Present	Absent
Present	300	51	53	180
Absent	187	166	76	-

We also consider matching error as in [23], where the Matching Error Study ([42]) was used in order to evaluate both the probability of missing a link in linkage procedure between Census and PES, and the missing linkage probability of linkage involving ALS, in absence of better quantitative information, under the assumption of no errors in the rematch. The results of the Matching Error Study for 1990 U.S. Census in St. Louis stratum are reported in Table 3.4 (see table 4 page 562 in [23]).

Ignoring the unresolved cases, Fienberg and Ding [23] claimed that the probability of missing a true link can be estimated as $1 - \hat{\alpha}_1 = 1 - \hat{\alpha}_2 = 9/(2667 + 9) = 0.3363\%$. Following the same reasoning we evaluate the probability of false link, $\hat{\beta}_1 = \hat{\beta}_2 = 7/(7 + 427) = 1.6129\%$.

Fienberg and Ding [23] examined various log-linear model with different dependency structure in order to better fit data in Table 3.12, and claimed that the [CP][PA] fit the data much better and produced an estimate $\hat{N} = 1599$. Applying their correction for missing links only, Fienberg and Ding [23] estimated $\hat{N}_{DF} = 1585$. With the error matrix specified in Table 3.10, we obtain $\hat{N}_{MDF} = 1680$. This value falls into both the confidence intervals of the previous estimates.

Example 6. The simulation study proposed in [17] extends Example 4 in Section 3.4.4, based on the fictitious population census data ([41]), from three sources. The first list represents enumerations from the Patient Register Data of the National Health Service (PRD); the second list of those from Customer Information System (CIS), which combines administrative data on tax and benefits; the third list of those from a decennial Census (CEN). The linking variables (names, dates of birth, addresses) in the datasets may be distorted by missing values and typos, in order to mimic real-life situations. The true match pattern and linking variables are available for evaluation of the results.

The simulation generated 500 replicate target populations of size 1000, independently and randomly sampled from the true data without replacement. For each replicate, three lists were randomly extracted from the PRD, CIS and CEN, according to the capture probabilities $\pi_1 = 0.65$, $\pi_2 = 0.53$ and $\pi_3 = 0.57$, respectively.

For each replication, the linkage was performed as explained in subsection 3.3.1: at the first step, PRD and CIS were linked and at the second step, the linked and unlinked cases resulting from the first step were linked with CEN. In both steps, the linkage variables were Name, Surname, Day, Month and Year of Birth, and the probabilistic record linkage models ([21]) were performed by means of the batch version of the software RELAIS ([47]).

TABLE 3.13
Distribution of the linkage error rates

Linkage Errors %	Min	Median	Mean	Max
First step				
$1-\alpha_1$	0.00	2.39	2.51	7.33
β_1	0.00	4.58	4.31	7.63
Second step				
$1-\alpha_2$	0.90	2.86	2.91	5.93
β_2	0.20	4.43	4.01	8.12

Table 3.13 summarizes the linkage results in terms of linkage error rates, reporting the probability of missing true matches $1 - \alpha$ and the probability of false link β as defined in Section 3.3.2 for both steps. The true values of α and β can be evaluated in light of the known true match pattern.

From each replication, we computed the naïve log-linear estimator and the adjusted MDF estimator described in Section 3.5.1. Now that the three

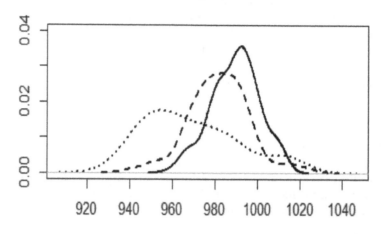

FIGURE 3.1
Simulated alternative estimates of the population size (N=1000).

lists were generated independently, the adopted log-linear model was the one assuming independence between the lists. The MDF estimator is computed using the true values of α and β obtained in each replication. The use of these true values allows us to compare the alternative estimators without the effects of linkage error estimation. The distributions of the two estimates over the 500 replications are shown in Figure 3.1, where MDF is designated as "Adjusted DCT" and the estimates obtained using the true counts without linkage errors are included for comparison.

The percentage relative errors of the estimators are summarized in Table 3.14, where the minimum, the first quartile, the median, the mean, the third quartile and the maximum of the Percentage Relative Error over the 500 replications are reported. Table 3.14 reports results for Naïve estimator; the Ding and Fienberg (DF) estimator, where the adjustment only accounts for missing links; the MDF estimator, adjusted for both missing and false link errors according to the matrix M_2 in Table 3.10; and the true estimator when the linkage is perfect. Table 3.14 reports both the estimates obtained through log-linear model and the estimates in closed form, derived by formula 3.5 in the presence of independency of the lists.

The results in Figure 3.1 and Table 3.14 demonstrate that the proposed adjustments reduce the bias of the Naïve estimator without side effects on the variance of the estimator, even if the bias is not entirely removed due to the approximate nature of the adjustments.

TABLE 3.14
Distribution of percentage relative error

	Percentage Relative Error					
Using the log-linear model						
Estimator	Min	Q1	Median	Mean	Q3	Max
Naive	-8.70	-4.90	-3.70	-3.19	-1.70	3.90
DF	-11.70	-8.11	-6.70	-6.67	-5.40	-1.60
MDF	-5.90	-2.70	-1.80	-1.75	-1.00	3.50
True	-4.80	-1.82	-1.00	-1.05	-0.20	2.30
From formula (3.5)						
Estimator	Min	Q1	Median	Mean	Q3	Max
Naive	-6.52	-2.09	0.03	0.60	2.94	15.42
DF	-12.30	-6.22	-4.40	-4.11	-2.27	7.5
MDF	-6.40	-1.20	1.10	1.36	3.42	16.00
True	-5.06	-0.95	0.56	0.83	2.31	10.79

3.6 Concluding remarks

The focus of this chapter is on the uncertainty due to record linkage in population-size estimation. The combination of multiple sources of data allows one to estimate the size of the population when each single source is affected by under-coverage of the target population. Traditionally this problem is dealt with by log-linear models and their extensions to address the violations of the model assumptions. We described several solutions to estimate the population size when the perfect linkage assumption is not met.

In this estimation framework, the linkage errors play a crucial role compared to the standard regression models. In regression analysis, the presence of false links attenuates the face-value estimates of the regression coefficient towards zero, whereas a missing link can be viewed as nonresponse and dealt with by the missing at random assumption, e.g., [12]. In population-size estimation, missing links affect the counts in the contingency Tables 3.1 and 3.2 where false links tend to bias the face-value population-size estimates downwards, while missing links cause bias upwards. Hence both need to be adjusted explicitly. Moreover, the impact of linkage errors on population-size estimation can be appreciable even when the size of the errors is small, particularly when the capture probabilities are close to one.

The adjusted methods for the two list cases have been originally proposed by [19] for the census post-enumeration survey case and extended by [16] to general frameworks. The three list cases has been firstly approached by [23] adjusting only the false links; the false links have been included by [17], which also describe linkage scenarios for multiple lists. The main results show that the adjusted estimators allow reducing the bias of the naïve estimator without unduly inflating the variance.

The more sophisticated adjustment methods depend on correct evaluation of both kinds of linkage errors. This is clearly seen in the real data application. Examples are given where the estimators' performances are assessed in both scenarios: linkage errors are estimated from the linkage model and the true linkage error values are available. Di Consiglio and Tuoto [17] provide a sensitivity analysis to show the effectiveness of the adjustments with respect to the misspecification of the linkage errors. Indeed, the evaluation of linkage errors is still an open issue. Tuoto [61] proposes a supervised learning method to predict both types of linkage errors, without relying on strong distribution assumptions, as in [5] . Alternatively, Chipperfield and Chambers [11] apply a bootstrap method to the actual linkage procedure to evaluate the mismatch probabilities. The proposals that consider the linkage errors in analyses on linked data are often based on a training set to assess linkage quality. Anyway, automatic probabilistic methods are necessary, particularly for detecting missing link errors.

A further improvement in adjusting for linkage errors could be achieved by allowing for varying probabilities of correct link and missing link. Indeed, the described adjustments assume constant linkage errors across the entire population. When this does not hold, the adjustment can still be applied in homogeneous strata with respect to linkage errors. The gain of the adjusted estimators in the presence of homogeneous strata compared to the use of average values of the errors over the entire population could be examined; this is an aspect to be tackled in future research.

There is also a need to improve the estimation methods given unbiased or nearly unbiased estimation of the linkage error probabilities. As seen in Example 6, the MDF estimator using the true error rates still has a slightly larger bias, compared to the estimator using the true data that is perfectly linked. We believe this may be at least partly due to the non-linear nature of the estimator in this case. But it is not entirely clear whether there is also a confounding effect with the heterogeneous linkage error probabilities.

Finally, the chapter describes shortly the Bayesian proposals by [39], [59], [60], [38], which adopts a unified framework for linkage and estimation. However, the lack of scalability is currently a serious drawback.

Bibliography

[1] Agresti, A. (1994). Simple capture-recapture models permitting unequal catchability and variable sampling effort. Biometrics 50, 494-500.

[2] Bailey, L. L., Kendall, W. L., and Church, D. R. (2008). Exploring extensions to multi-state models with multiple unobservable states. In Envi-

ronmental and Ecological Statistics Vol. 3, D. L. Thomson, E. G. Cooch, and M. J. Conroy (eds.), 693–709. New York: Springer.

[3] Ball, P., W. Betts, F.J. Scheuren, J. Dudukovich, and J. Asher (2002). Killings and Refuge Flow in Kosovo, March – June 1999, American Academy for the Advancement of Science, January 3.

[4] Bartolucci, F. and Forcina, A. (2006). A class of latent marginal models for capture-recapture data with continuous covariates. Journal of the American Statistical Association, 101, 786-794.

[5] Belin, T. R. and Rubin, D. B. (1995). A Method for Calibrating False-Match Rates in Record Linkage. Journal of the American Statistical Association, 90, 694-707.

[6] Bock, R. D. (1975). Multivariate statistical methods in behavioral research. McGraw-Hill, New York.

[7] Böhning, D., Van der Heijden, P.G.M. and Bunge, J. (2017). Capture-Recapture Methods for the Social and Medical Sciences. Chapman and Hall/CRC.

[8] Chambers, R. (2009). Regression Analysis Of Probability-Linked Data. Official Statistics Research Series, Vol 4, 2009. ISSN 1177-5017; ISBN 978-0-478-31569-1.

[9] Chao, A. (2001). An overview of closed Capture-Recapture Models. Journal of Agricultural, Biological, and Environmental Statistics, 6, 158-175.

[10] Chen, Z. and Kuo, L. (2001). A note on the estimation of the multinomial Logit Model with Random effects. The American Statistician, 55, 89-95.

[11] Chipperfield, J. O. and Chambers, R. L. (2015). Using the Bootstrap to Account for Linkage Errors when Analysing Probabilistically Linked Categorical Data, Journal of Official Statistics, 31(3), 397-414.

[12] Chipperfield, J. O., Bishop, G. R. and Campbell, P. (2011). Maximum likelihood estimation for contingency tables and logistic regression with incorrectly linked data Survey Methodology, June 2011 13, Vol. 37, No. 1, pp. 13-24.

[13] Copas, J. B. and Hilton, F. J. (1990). Record linkage: Statistical models for matching computer records. J. Roy. Statist. Soc. Ser. A, 153, 287–320.

[14] Cormack, R. M. (1989). Log-linear models for capture-recapture. Biometrics, 45, 395-413.

[15] Coull, B. A. and Agresti, A. (1999). The Use of Mixed Logit Models to Reflect Heterogeneity in Capture-Recapture Studies. Biometrics, 55, 294-301.

[16] Di Consiglio, L. and Tuoto, T. (2015). Coverage Evaluation on Probabilistically Linked Data, Journal of Official Statistics, Vol. 31, No. 3, pp. 415–429.

[17] Di Consiglio, L. and Tuoto, T. (2018). Population size estimation and linkage errors: the multiple lists case, Journal of Official Statistics, 34(4), 889-908.

[18] Dempster, A. P., Laird, N. M. and Rubin, D. B. (1977). Maximum Likelihood From Incomplete Data via the EM Algorithm (with comments). Journal of the Royal Statistical Society, Ser. B, 39, 1-37.

[19] Ding, Y. and Fienberg, S.E. (1994). Dual system estimation of Census undercount in the presence of matching error. Survey Methodology, 20, 149-158.

[20] Farcomeni, A. and Tardella, L. (2009). Reference Bayesian methods for recapture models with heterogeneity. Test, May 2010, Volume 19, Issue 1, pp. 187–208.

[21] Fellegi, I. P. and Sunter, A. B. (1969). A Theory for Record Linkage. Journal of the American Statistical Association, 64, 1183-1210.

[22] Fienberg, S. E. (1972). The multiple recapture census for closed populations and incomplete 2k contingency tables. Biometrika 59, 3, 591-603.

[23] Fienberg, S.E. and Ding, Y. (1996). Multiple sample estimation of population and census undercount in the presence of matching error. In Proceedings of 1994 Annual research conference and CASIC technologies Interchange, Bureau of Census, US.

[24] Fienberg, S. E. (1991). An Adjusted Census in 1990? Commerce Says "No", Chance, 4, 44-52.

[25] Fienberg, S.E. and D. Manrique-Vallier. (2009). Integrated Methodology for Multiple Systems Estimation and Record Linkage Using a Missing Data Formulation. Advances in Statistical Analysis 93, 49–60.

[26] Fortini, M., Liseo, B., Nuccitelli, A. and Scanu, M. (2001). On Bayesian Record Linkage. Research in Official Statistics, 4, Vol.1, 185-198.

[27] Freedman, D.A. (1991). Policy Forum: Adjusting the 1990 Census, Science, 252, 1, 233-236.

[28] Ghosh, S.K., Norris, J.L. (2004). Bayesian Capture-Recapture Analysis and Model Selection Allowing for Heterogeneity and Behavioral Effects. NCSU Institute of Statistics, Mimeo Series 2562, pp. 1-27.

[29] Herzog, T., F. Scheuren, and W. Winkler. (2007). Data Quality and Record Linkage Techniques. New York: Springer-Verlag.

[30] Hogan, H. (1992). The 1990 Post-Enumeration Survey: an Overview, The American Statistician, 45, 261-269.

[31] IWGDMF - International Working Group for Disease Monitoring and Forecasting (1995). Capture-recapture and multiple-record systems estimation I: History and theoretical development. American Journal of Epidemiology, 142, 1047-1058.

[32] Isaki, C. T. and Schultz, L. K. (1987). The effects of correlation and matching error on dual system estimation, Communications in Statistics - Theory and Methods, 16, 2405-2427.

[33] Jaro, M. (1989). Advances in record linkage methodology as applied to matching the 1985 test census of Tampa, Florida. Journal of American Statistical Association, 84, 414-420.

[34] Larsen, M. D. (1996). Bayesian Approaches to Finite Mixture Models, Ph.D. dissertation, Harvard University, Dept. of Statistics.

[35] Larsen, M. D. and Rubin, D. B. (2001). Iterative Automated Record Linkage Using Mixture Models. Journal of the American Statistical Association, 96, 32-41.

[36] Larsen, M. (2005). Advances in record linkage theory: hierarchical Bayesian record linkage. Proceedings of the Section on Survey Research Methods, American Statistical Association, pp. 3277-3283.

[37] Lincoln, F. C. (1930). Calculating Waterfowl Abundance on the Basis of Banding Returns. United States Department of Agriculture Circular, 118, 1–4.

[38] Link, W.A., Yoshizaki J., Bailey L.L., Pollok K.H. (2010). Uncovering a latent multinomial: Analysis of Mark-recapture data with misidentification, Biometrics, 66, 178-185.

[39] Liseo, B., Tancredi, A. (2011). Bayesian estimation of population size via linkage of multivariate normal data sets. Journal of Official Statistics, Vol. 27 No. 3, pp. 491-505.

[40] McFadden, D. (1974). Conditional logit analysis of qualitative choice behavior. In P. Zarembka (ed.), Frontiers in Econometrics, 105–142.

[41] McLeod, P., Heasman, D. and Forbes, I. (2011). Simulated data for the on the job training. Essnet DI available at http://www.cros-portal.eu/content/job-training.

[42] Murly, M.H., Dajani, A., Biemer, P. (1989). The Matching Error Study for the 1988 Dress Rehearsal Proceedings of the Section on Survey Research Methods, ASA, 704-709.

[43] Otis, D.L., Burnham, K.P., White, G.C. and Anderson, D.R. (1978). Statistical inference from capture data on closed animal populations. Wildlife Monographs, 62.

[44] Parag and Domingos, P. (2004). Multi-Relational Record Linkage In Proceedings of the KDD-2004 Workshop on Multi-Relational Data Mining.

[45] Petersen, C.G.J. (1896). The yearly immigration of young plaice into the Limfiord from the German Sea. Report of the Danish Biological Station, 6, 5–84.

[46] Pollock, K.H., Nichols, J.D., Brownie, C. and Hines, J.E. (1990). Statistical inference for capture-recapture experiments. Wildlife monographs, 107.

[47] RELAIS, (2015). User's guide version 3.0, available at https://joinup.ec.europa.eu/software/relais/asset_release/relais-221.

[48] Robert, C. P. and Casella, G. (2004). Monte Carlo Statistical Methods, 2nd ed. Springer, New York.

[49] Sadinle, M. and S.E. Fienberg. 2013. A Generalized Fellegi-Sunter Framework for Multiple Record Linkage With Application to Homicide Record Systems. JASA, 108: 385–397.

[50] Sadinle, M., R. Hall, and S.E. Fienberg. (2011). Approaches to Multiple Record Linkage. In Proceedings of the ISI World Statistical Congress, 21-26 August 2011, Dublin, 1064–1071.

[51] Sanathanan, L. (1972). Estimating the size of a multinomial population. Annals of Mathematical Statistics, 43, 142-152.

[52] Seber, G.A.F. (1982). The estimation of animal abundance and related parameters. Second edition. Macmillan, New York, USA.

[53] Seber, G.A.F. (1986). A review of estimating animal abundance. Biometrics 42: 267-292.

[54] Sekar, C.C. and W.E. Deming. (1949). On a Method of Estimating Birth and Death Rates and the Extent of Registration, J. Amer. Stat. Assoc., Vol. 44, pages 101-115.

[55] Steorts, R., R. Hall, and S.E. Fienberg. (2014). SMERED: A Bayesian Approach to Graphical Record Linkage and De-duplication. Journal of Machine Learning Research 33: 922–930.

[56] Steorts, R., R. Hall, and S.E. Fienberg. (2015). A Bayesian Approach to Graphical Record Linkage and De-duplication. Journal of the American Statistical Association. URL http://arxiv.org/abs/1312.4645

[57] Steorts, R. (2015). Entity Resolution with Empirically Motivated Priors, Bayesian Analysis 10, Number 4, pp. 849–875.

[58] Stevick P.T., Palsboll P.J., Smith T.D., Bravington M.V., and Hammond P.S. (2001). Errors in identification using natural markings: rates, sources, and effects on capture–recapture estimates of abundance. Can. J. Fish. Aquat. Sci. 58: 1861–1870.

[59] Tancredi, A. and Liseo, B. (2011). A hierarchical Bayesian approach to record linkage and population size problems. Annals of Applied Statistics, 5(2B): 1553–1585. MR2849786. doi: http://dx.doi.org/10.1214/10-AOAS447. 851, 858, 860.

[60] Tancredi, A., Steorts, R., Liseo, B. (2018). A unfied framework for data deduplication and population size estimation. To appear.

[61] Tuoto, T. (2016). New proposal for linkage error estimation. Statistical Journal of the IAOS, Vol. 32, no. 2, pp. 1-8.

[62] Ventura, S. and R. Nugent. (2014). Hierarchical Clustering with Distributions of Distances for Large-Scale Record Linkage. In Privacy in Statistical Databases, edited by J. Domingo-Ferrer, pp. 283–298. Berlin: Springer Link. Lecture Notes in Computer Science, Vol. 8744.

[63] Wolter, K.M. (1986). Some coverage error models for census data. Journal of the American Statistical Association, 81, 338-346.

[64] Wolter, K.M. (1991). Policy Forum: Accounting for America's Uncounted and Miscounted, Science, 253, 12-15.

[65] Zhang, L.-C. and Dunne, J. (2017). Trimmed dual system estimation. Chapter 17, pp. 239-259, in Capture-Recapture Methods for the Social and Medical Sciences, edited by D. Bohning, P.G.M. Van der Heijden, J. Bunge, Chapman and Hall/CRC.

[66] Zhang, L.-C. (2018). A note on dual system population size estimator. Journal of Official Statistics. To appear.

[67] Zwane, E. and Van Der Heijden, P. (2005). Population estimation using the multiple system estimator in the presence of continuous covariates. Statistical Modelling, 5, 39-52.

[68] De Wolf, P.P., Van Der Laan, J., Zult, D. (2018). Joining correction methods for linkage error in capture-recapture, 45, Discussion paper, Statistics Netherlands, The Hague/Heerlen.

4

An overview on uncertainty and estimation in statistical matching

Pier Luigi Conti
Sapienza University of Rome, Rome, Italy

Daniela Marella
Università Roma TRE, Rome, Italy

Mauro Scanu
Istat, National Institute of Statistics, Rome, Italy

CONTENTS

4.1 Introduction

The necessity to offer quick answers to the growing informative needs of today's society can be accomplished by the joint use of two or more sample surveys drawn from the same population. The interesting inferential aspect in their joint use is that, for most sample designs, the probability that two samples observe a common subset of units is negligible; hence, it is impossible to reconstruct the joint observations of pairs of variables, one observed only in one sample and not in the other, and the opposite for the other variable. This problem is often called statistical matching, and it is of increasing interest (see [18]).

One of the long-debated applications of statistical matching is in the framework of household-level data on income and expenditure, widely used by policy makers to provide insights into a number of areas (see [34], [43]; see also [14] and references therein). First of all, the analysis of the impact of fiscal policy proposals often requires information on household expenditures and income. This is usually achieved through micro-simulation models which simulate the short-term or long-term spending/saving behavior of households in case, for instance, of tax reforms or changes in the pension system.

Next, most policy initiatives aimed at improving living standards tend to measure poverty using income or consumption. However, both income and consumption are not sufficient measures of poverty when considered separately. A better approach is to use both simultaneously, particularly if one considers poverty in terms of achieved standards of living. Because of this, consumption of goods and services is arguably a more important determinant of economic well-being than income alone. All poverty and inequality measures are generally functions of quantile estimates of the income and expenditure distributions.

A major drawback is that most countries do not have single micro-data sources including high-quality disaggregated joint data on income and expenditures for two reasons: avoiding response burden and ensuring high quality answers to the questionnaires. As a consequence, since asking detailed questions on income and consumption in the same survey can be problematic, surveys tend to specialize in just one of the two topics.

In Italy this problem has been tackled by using reliable information on household income provided by the EU-SILC (Statistics on Income and Living Conditions) survey, which is annual and conducted by ISTAT (Italian National Institute of Statistics). On the other hand, reliable and precise information on consumption expenses are provided by the ISTAT sample survey (Household Budget Survey, HBS for short), which is conducted every year. Both samples are selected, according to a multistage, complex sampling design. ISTAT databases provide relevant information *separately* for income and consumption expenses. However, they do not contain *any* joint observation of household *income and consumption expenses*. Then a natural question arises: *Is it possible, to some extent, to estimate the joint distribution of income and consumption expenses on the basis of the two above databases?* This is, in short, the problem of *statistical matching*. The above question is essentially equivalent to this: *Is it possible, to some extent, to (re)construct a unique database containing* both income and consumption expenses, without resorting to an ad hoc, expensive, survey?

This problem is generally overcome with statistical matching of independent sample surveys referred to the same target population (see [23]; see also [21] and references therein).

Although most of the applications have been on income and expenditures, this problem is going to become ubiquitous in official statistics. The growing need to cut costs and have timely results have brought many National Statistical Institutes to adopt modernization problems that include, among other objectives, the coexistence of data coming from different sources in a unique register. These registers multiply the possibility to extend the use of statistical matching for all those pairs of variables for which directly observed joint data do not exist and cannot be recreated linking data that belongs to the same statistical unit. Furthermore, additional questions arise, as for the possible use of statistical matching techniques when big data are available ([20], [40]).

There is an extensive bibliography on this subject, that dates back to the late 1960s. It seems that the two prominent problems in this setting are the absence of joint observations on the pair of variables of interest, that induces a specific form of uncertainty; how to use jointly, for statistical matching purposes, two samples that can be possibly drawn according to different complex survey designs. This chapter reviews the solutions available for these two problems, respectively in 4.2 and 4.4. Finally, some lines on the connections between the statistical matching problem and a research problem developed independently (ecological inference) is discussed in Section 4.5.

4.2 Statistical matching problem: notations and technicalities

Let U be a finite population of N units, labeled by integers $1, \ldots, N$, and let (X, Y, Z) be three variables of interest. Denote further by x_i, y_i, z_i the values of X, Y, Z, respectively, for unit $i\ (= 1, \ldots, N)$, and by $\mathbf{x}_N, \mathbf{y}_N, \mathbf{z}_N$ the sequences:

$$\mathbf{x}_N = (x_1 \ldots, x_N), \ \mathbf{y}_N = (y_1 \ldots, y_N), \ \mathbf{z}_N = (z_1 \ldots, z_N). \tag{4.1}$$

The knowledge of (X, Y, Z) over the whole population is essentially equivalent to the knowledge of the joint population distribution function (p.d.f., for short)

$$H_N(x, y, z) = \frac{1}{N} \sum_{i=1}^{N} I_{(x_i \le x)} I_{(y_i \le y)} I_{(z_i \le z)}, \ x, y, z \in \mathbb{R} \tag{4.2}$$

where $I_{(\phi)}$ is the indicator of event ϕ. Let

$$F_N(x, y) = \frac{1}{N} \sum_{i=1}^{N} I_{(x_i \leq x)} I_{(y_i \leq y)}, \tag{4.3}$$

$$G_N(x, z) = \frac{1}{N} \sum_{i=1}^{N} I_{(x_i \leq x)} I_{(z_i \leq z)}, \tag{4.4}$$

$$Q_N(x) = \frac{1}{N} \sum_{i=1}^{N} I_{(x_i \leq x)} \tag{4.5}$$

be the joint p.d.f.s of (X, Y), (X, Z) and the marginal p.d.f. of X, respectively, and let

$$p_N(x) = Q_N(x) - Q_N(x^-) = \frac{1}{N} \sum_{i=1}^{N} I_{(x_i = x)} \tag{4.6}$$

be the proportion of population units such that $X = x$. Define further the conditional p.d.f.s

$$F_N(y|x) = \frac{F_N(x, y)}{p_N(x)}, \tag{4.7}$$

$$G_N(z|x) = \frac{G_N(x, z)}{p_N(x)}, \tag{4.8}$$

$$H_N(y, z|x) = \frac{H_N(x, y, z)}{p_N(x)} \tag{4.9}$$

that can be arbitrarily defined whenever $p_N(x) \neq 0$.

Partial sampling information about the three characters X, Y, Z is frequently available, for example, in the sample surveys regularly collected by National Statistical Offices. We assume that:

1. a sample s_A of n_A units is selected *via* a complex sample design from the finite population U, and the values (x_i, y_i), $i \in s_A$, are observed;

2. a sample s_B of n_B units is selected, independently of s_A, *via* a complex sample design from the finite population U, and the values (x_i, z_i), $i \in s_B$, are observed;

3. the two samples s_A, s_B are independent.

Since, at least in real social surveys, the probability of selecting a common unit in s_A and s_B is essentially negligible (see [42]), it can be assumed that the two samples s_A, s_B do not overlap. Roughly speaking, the observational mechanism is such that (i) only the variables (X, Y) are observed in s_A, and (ii) only the variables (X, Z) are observed in s_B. The variable X is *common* to the samples s_A, s_B, and plays the role of matching variable ([38] calls these matching variables as the glue used in order to reconstruct the broken Y and Z observations). Hence, the variables X, Y, Z are not jointly observed.

4.3 The joint distribution of variables not jointly observed: estimation and uncertainty

Statistical matching aims at combining information in s_A and s_B. More precisely, at a "macro" level, the goal of statistical matching consists in using the sample data

$$\{(x_i, y_i); i \in s_A\}, \ \{(x_i, z_i); i \in s_B\} \tag{4.10}$$

to estimate the joint p.d.f. (4.2).

Since X, Y, Z are not jointly observed, unless special assumptions are made, the statistical model for the joint distribution of (X, Y, Z) is usually *unidentifiable*. That is, the joint p.d.f. (4.2) cannot be estimated on the basis of the available sample information. Formally speaking, the estimation of the joint p.d.f. of (X, Y, Z) is essentially equivalent to the estimation of (i) the marginal distribution of X; (ii) the joint distribution of (Y, Z) conditionally on X.

Since there are no joint observations of Y and Z (neither marginally, nor conditionally on X), the only quantity that *cannot* be estimated from sample data is $H_N(y, z|x)$ in (4.9). This lack of identifiability is the cause of "intrinsic" *uncertainty* about the statistical model of (X, Y, Z). Even if the sample sizes n_A, n_B are large, one can only expect reasonably accurate estimates of the bivariate distributions of (X, Y) and (X, Z), respectively, but *not* of the whole distribution of (X, Y, Z). That is, the statistical model would remain *unidentifiable*.

An identifiable model for the data at hand is the one that assumes the independence of Y and Z given X. This assumption is the well-known *Conditional Independence Assumption* (CIA, for short). Under the CIA the joint p.d.f. (4.2) can be factorized as follows

$$H_N(y, z|x) = F_N(y|x)G_N(z|x). \tag{4.11}$$

Appropriateness of CIA is discussed in several papers. We cite, among others, [43] and [41].

Then a natural question arises: when the CIA does not hold, what can we say about $H_N(y, z|x)$?

In the absence of any assumptions or prior information about the dependence between Y and Z and if only the p.d.f.s (4.3)–(4.5) were known, then one could only say that

$$\max(0, F_N(y|x) + G_N(z|x) - 1) \le H_N(y, z|x) \le \min(F_N(y|x), G_N(z|x)). \tag{4.12}$$

The bounds in (4.12) are the well-known *Fréchet bounds*. They represent the minimal and maximal pointwise values $H_N(y, z|x)$ can take, respectively. We stress that the lower Fréchet bound corresponds to the maximal negative association between Y and Z, given X; this comes true if and only if (iff) Y is a strictly decreasing function of Z (and vice versa), given X. Similarly, the upper Fréchet bound corresponds to the maximal positive association between Y and Z, given X; this comes true iff Y is a strictly increasing function of Z (and vice versa), given X.

Example 1. Multinormal distribution - This example has been extensively studied, since [2]. Let X, Y, Z be jointly multinormally distributed, with mean vector and covariance matrix equal to

$$\mu = \begin{bmatrix} \mu_x \\ \mu_y \\ \mu_z \end{bmatrix}, \quad \Sigma = \begin{bmatrix} \sigma_x^2 & \sigma_{xy} & \sigma_{xz} \\ \sigma_{xy} & \sigma_y^2 & \sigma_{yz} \\ \sigma_{xz} & \sigma_{yz} & \sigma_z^2 \end{bmatrix} \tag{4.13}$$

respectively. Conditionally on X, (Y, Z) do have joint bivariate normal distribution with mean vectors and covariance matrices easily obtained from (4.13), given by

$$\mu_{yz|x} = \begin{bmatrix} \mu_y + \beta_{y/x}(x - \mu_x) \\ \mu_z + \beta_{z/x}(x - \mu_x) \end{bmatrix}, \tag{4.14}$$

$$\Sigma_{yz|x} = \begin{bmatrix} \sigma_y^2(1 - \rho_{xy}^2) & \sigma_y\sigma_z(\rho_{yx} - \rho_{xy}\rho_{xz}) \\ \sigma_y\sigma_z(\rho_{yx} - \rho_{xy}\rho_{xz}) & \sigma_z^2(1 - \rho_{xz}^2) \end{bmatrix} \tag{4.15}$$

Let $\rho_{xy}, \rho_{xz}, \rho_{yz}$ be the correlation coefficients between (X,Y), (X,Z), (Y,Z), respectively. The only unidentified parameter is ρ_{yz}, the correlation coefficient between Y and Z.

From (4.15), it is immediate to see that, given ρ_{xy} and ρ_{xz}, ρ_{yz} ranges in the interval

$$\left[\rho_{xy}\rho_{xz} - \sqrt{(1 - \rho_{xy}^2)(1 - \rho_{xz}^2)}, \; \rho_{xy}\rho_{xz} + \sqrt{(1 - \rho_{xy}^2)(1 - \rho_{xz}^2)} \right]. \tag{4.16}$$

All the values in (4.16) are equally *plausible*. Note that, under the CIA, the parameter ρ_{yz} is located at the midpoint of the interval (4.16), as remarked in [32].

From the Slepian inequality [45], the joint d.f. of Y, Z given X is an increasing function of the correlation coefficient between Y and Z given X:

$$\rho_{yz|x} = \frac{\rho_{yz} - \rho_{xy}\rho_{xz}}{\sqrt{(1 - \rho_{xy}^2)(1 - \rho_{xz}^2)}}$$

i.e., it turns out to be a monotone function of ρ_{yz}. In other words, the conditional d.f.s $H_N(y, z|x)$ are totally ordered on the basis of ρ_{yz}. The case $\rho_{yz} = \rho_{xy}\rho_{xz} - \sqrt{(1 - \rho_{xy}^2)(1 - \rho_{xz}^2)}$ corresponds to $\rho_{yz|x} = -1$, so that conditionally on X, Y is a linear decreasing function of Z (and vice versa). As a consequence, when the only available information are the data the

Fréchet lower bound is $\max(0, F_N(y|x) + G_N(z|x) - 1)$. Similarly, the case $\rho_{yz} = \rho_{xy}\rho_{xz} + \sqrt{(1 - \rho_{xy}^2)(1 - \rho_{xz}^2)}$ corresponds to $\rho_{yz|x} = 1$, i.e., to the Fréchet upper bound $\min(F_N(y|x), G_N(z|x))$.

According to the previous example, it is possible to say that generally the main consequence of the lack of identifiability in a parametric setting is that some parameters of the model cannot be estimated on the basis of the available sample information. For instance, instead of parameter point estimates, one can only reasonably construct sets of *plausible estimates*, compatible with what can be actually estimated. These sets (usually intervals) formally provide a representation of uncertainty about the model parameters. A unified framework for the parametric and non-parametric approaches in analysing uncertainty in statistical matching is in [16].

Fréchet bounds (4.12) can be improved when extra-sample information is available. In statistical practice, a kind of extra-sample information frequently available consists of *logical constraints*, namely in restrictions on the support of the variables (Y, Z) given X. A sufficiently general constraint has the following form

$$a_x \le f_x(y, z) \le b_x \tag{4.17}$$

given $X = x$, where $f_x(y, z)$ is an increasing function of y for fixed z and a decreasing function of z for fixed y [19].

In practice, the constraint (4.17) means that for each unit i of the population $\mathcal{U}_N$ the values (x_i, y_i, z_i) must satisfy the pair of inequalities $a_{x_i} \le f_{x_i}(y_i, z_i) \le b_{x_i}$. Roughly speaking, the constraint (4.17) modifies the support of the conditional p.d.f. $H_N(y, z|x)$, that becomes

$$\{(y_i, z_i): a_{x_i} \le f_{x_i}(y_i, z_i) \le b_{x_i} \text{ and } x_i = x\}.$$

Under the constraint (4.17), the Fréchet bounds (4.12) reduce to

$$K_{N-}^x(y, z) \le H_N(y, z|x) \le K_{N+}^x(y, z) \tag{4.18}$$

where (using $\wedge$ for the minimum between two numbers)

$$
\begin{aligned}
K_{N-}^x(y, z) &= \max(0, G_N(z|x) \wedge G_N(\gamma_y(a_x)|x) + F_N(y|x) \wedge F_N(\delta_z(b_x)|x) - 1, \\
&\quad F_N(y|x) + G_N(z|x) - 1) \tag{4.19} \\
K_{N+}^x(y, z) &= \min(G_N(z|x), G_N(\gamma_y(a_x)|x), F_N(y|x), F_N(\delta_z(b_x)|x)) \tag{4.20}
\end{aligned}
$$

and $\gamma_y(\cdot)$, $\delta_z(\cdot)$ being the inverse functions of $f_x(y, z)$ for fixed y and z, respectively. Since

$$
\begin{aligned}
K_{N-}^x(y, z) &\ge \max(0, F_N(y|x) + G_N(z|x) - 1), \\
K_{N+}^x(y, z) &\le \min(F_N(y|x), G_N(z|x)),
\end{aligned}
$$

the bounds (4.18) are actually an improvement of the unconstrained Fréchet bounds (4.12).

Example 2. Assume that there exist constants a_x, b_x and let $f_x(Y,Z) = Y/Z$, then the constraint assumes the form $a_x \leq Y/Z \leq b_x$. For instance, in household surveys, X is a household socio-economic character (i.e., the number of household members), Y the household consumption and Z the household income. Then using techniques of national accounting, it is possible to produce fairly reasonable lower and upper bounds of the average propensity to consume, namely of the ratio between consumption expenditure and income, for each household size. As a consequence, for each household size, lower and upper bounds for the consumption expenditure for a given income can be computed. Such a kind of constraint is analyzed in [14], where the statistical matching between the Banca d'Italia Survey on Household Income and Wealth (SHIW, for short) and the Italian National Statistical (Istat) Household Budget Survey (HBS, henceforth) has been performed.

The constraint introduced above means that, for each given x, the support of the conditional p.d.f. $H_N(y, z|x)$ is in between the two straight lines $z = y/b_x$ and $z = y/a_z$, i.e., is (a subset either proper or improper of) the cone

$$\{(y_i, z_i): y_i \geq 0, y_i/b_x \leq z_i \leq y_i/a_x\}. \tag{4.21}$$

Let $\gamma_y(a_x) = y/a_x$, $\delta_z(b_x) = b_x z$ from the results (4.19), (4.20) we obtain the following Fréchet bounds

$$K_{N-}^x(y, z) = \max(0, G_N(z|x) \wedge G_N\left(\frac{y}{a_x}|x\right) + F_N(y|x) \wedge F_N(b_x z|x) - 1,$$

$$F_N(y|x) + G_N(z|x) - 1) \tag{4.22}$$

$$K_{N+}^x(y, z) = \min\left(G_N(z|x), G_N\left(\frac{y}{a_x}|x\right), F_N(y|x), F_N(b_x z|x)\right) \tag{4.23}$$

Another example of the constraint $a_x \leq Y/Z \leq b_x$ happens in business surveys, X could be the type of activity, Y the total sales and Z the number of employees.

Example 3. Another kind constraint frequently occurring in practice is $Y \geq Z$, given X. For instance, this is the case of [34] where Y plays the role of total income and Z plays the role of income subject to taxation. The constraint $Y \geq Z$ means that, for each given x, the support of the conditional p.d.f. $H_N(y, z|x)$ becomes

$$\{(y_i, z_i): y_i \geq z_i\}. \tag{4.24}$$

then it is (a subset of) the half-plane below the straight line $z = y$.

Set $f_x(y, z) = y/z$ with $a_x = 1$ and $b_x \to \infty$, from the results (4.19), (4.20) we obtain the following new Fréchet bounds

$$
\begin{aligned}
K_{N-}^x(y, z) &= \max(0, G_N(z|x) \wedge G_N(y|x) + F_N(y|x) - 1, \\
&\quad F_N(y|x) + G_N(z|x) - 1) \quad &(4.25) \\
K_{N+}^x(y, z) &= \min(F_N(y|x), G_N(z|x), G_N(y|x)) \quad &(4.26)
\end{aligned}
$$

Note that the notion of uncertainty can be used in different tasks. D'Orazio et al. [22] use uncertainty in order to choose the best set of matching variables among those available in both datasets to match. Others, as [33], try to give alternative approaches for choosing parameter values in unidentifiable problems, through minimal inference. In this paper, we will mainly use uncertainty as a representation of what can be said of a joint distribution of a pair of variables for which there do not exist joint observed values, as well as for the representation of the matching error when one choice is definitely made (see Section 4.3.1).

4.3.1 Matching error

Since only joint observations of (X, Y) and (X, Z) are available, on the basis of sample data it is only possible to estimate $Q_N(x)$, $F_N(y|x)$, $G_N(z|x)$. Even when the conditional p.d.f.s $F_N(y|x)$ and $G_N(z|x)$ are completely known, the lack of joint observations on the variables (X, Y, Z) is the cause of uncertainty on $H_N(y, z|x)$. That is, the available information is unable to discriminate among a set of *plausible* joint distributions for (X, Y, Z).

As far as $H_N(y, z|x)$ is concerned, unless special assumptions are made, the only thing we can say is that it lies in the set of all *plausible* joint p.d.f.s defined as follow

$$
\begin{aligned}
\mathcal{H}_N^x &= \{H_N(y, z|x) : H_N(y, \infty|x) = F_N(y|x), \\
&\quad H_N(\infty, z|x) = G_N(z|x), a_x \leq f_x(y, z) \leq b_x\} \quad (4.27)
\end{aligned}
$$

The set $\mathcal{H}_N^x$ contains all joint p.d.f.s of $(Y, Z)|X$ having marginals $F_N(y|x)$ and $G_N(z|x)$ and satisfying the imposed logical constraint.

Then every bivariate distribution in the class (4.27) is a candidate to be the joint distribution of Y and Z given X. From now on, each bivariate d.f. in the class (4.27) will be called a *matching distribution* for Y and Z, given X.

The statistical matching problem essentially consists in choosing a matching distribution, i.e., a d.f. in the class $\mathcal{H}_N^x$.

A *matching procedure* is essentially a procedure to choose a d.f. in the class (4.27), or better, in a class defined as (4.27) but with marginal d.f.s estimated on the basis of sample data.

Suppose $H_N^*(y, z|x) \in \mathcal{H}_N^x$ is chosen as a matching distribution for (Y, Z) given X, but that the "true" d.f. is $H_N(y, z|x)$ (again in $\mathcal{H}_N^x$, of course). The discrepancy between the chosen $H_N^*(y, z|x)$ and the actual $H_N(y, z|x)$ is the *matching error*. The notion of matching error is of basic importance in validating matching procedures, because the smaller the matching error, the better the matching procedure.

The most favourable case for matching error occurs under CIA since the class (4.27) collapses into a *single* d.f. and from (4.11) in order to construct a consistent estimate of $H_N(y, z|x)$ it is enough to construct consistent estimates of $F_N(y|x)$ and $G_N(z|x)$. In this case, the matching error becomes negligible as the sizes of s_A and s_B increase.

The matching error produced by a class of non-parametric imputation procedures in the CIA simplified context and for *i.i.d* observations is evaluated in [30]. This class is based on the k Nearest Neighbour (kNN) non-parametric estimation of the regression function of Z on X in s_B, both with fixed and variable number of donors k, and includes some of the most popular non-parametric imputation procedures such as distance and random hot deck. The asymptotic properties of the imputation procedures are formally analyzed, and then studied by simulation. In [15] we go further by introducing new non-parametric matching techniques based on local linear regression whose performance is measured by the matching error.

Unfortunately, in many cases of practical importance, the CIA is not reasonable. In this case, the class (4.27) does not reduce to a single point, and the matching error *cannot* become negligible, even if the sizes of s_A and s_B increase.

In spite of this drawback, the study of the matching error is still of primary importance, because a "small" matching error means that the chosen matching distribution $H_N^*(y, z|x)$ is "close" to the actual $H_N(y, z|x)$, and hence replacing $H_N(y, z|x)$ by $H_N^*(y, z|x)$ does not produce a "large" error.

From now on, in order to assess the accuracy of the matching distribution $H_N^*(y, z|x)$ as estimator of the true p.d.f. $H_N(y, z|x)$, as a *matching error measure* (conditionally on X) we will use the following:

$$ME_x(H_N^*, H_N) = \int_{\mathbb{R}^2} \left| H_N^*(y, z|x) - H_N(y, z|x) \right| d[F_N(y|x)G_N(z|x)]. \quad (4.28)$$

With a similar reasoning, as an unconditional measure of matching error we can consider the following:

$$\begin{aligned} ME(H_N^*, H_N) &= \int_{\mathbb{R}} ME_x(H_N^*, H_N) \, dQ_N(x) \\ &= \int_{\mathbb{R}} \left\{ \int_{\mathbb{R}^2} \left| H_N^*(y, z|x) - H_N(y, z|x) \right| \right. \\ &\qquad \left. \times d[F_N(y|x)G_N(z|x)] \right\} dQ_N(x). \quad (4.29) \end{aligned}$$

4.3.2 Bounding the matching error via measures of uncertainty

This section is devoted to the notion of uncertainty in the statistical matching framework. This kind of uncertainty is due to the absence of joint information on the variables of interest, and is relevant when the statistical model for the joint p.d.f. is not identifiable.

A measure of uncertainty in statistical matching is important because it is related to the matching error, and consequently gives information on how reliable is the estimate of a *plausible* matching distribution. See [19] for technical details.

Note that, if on one hand the notion of matching error is crucial to validate a statistical matching procedure, on the other hand the sample data in s_A, s_B do not contain enough information to estimate the matching error (4.28). However, this drawback can be overcome using the analysis in Section 4.3.1. From the inequality

$$\left| H_N^*(y, z|x) - H_N(y, z|x) \right| \leq K_{N+}^x(y, z) - K_{N-}^x(y, z) \ \forall \, x, y, z$$

it is immediate to see that

$$ME_x(H_N^*, H_N) \leq \Delta^x(F_N, G_N) \ \forall \, x \tag{4.30}$$

where

$$\Delta^x(F_N, G_N) = \int_{\mathbb{R}^2} \left(K_{N+}^x(y, z) - K_{N-}^x(y, z) \right) d[F_N(y|x)G_N(z|x)]$$

$$= \frac{1}{N^2 p_N(x)^2} \sum_{i=1}^N \sum_{j=1}^N \left(K_{N+}^x(y_i, z_j) - K_{N-}^x(y_i, z_j) \right) I_{(x_i=x)} I_{(x_j=x)}. \tag{4.31}$$

The quantity $\Delta^x(F_N, G_N)$ in (4.31) represents, in a sense, a measure of the "size" of the class (4.27).

The smaller $\Delta^x(F_N, G_N)$, the smaller the matching error, the closer the matching distribution to the actual distribution of Y and Z given X. Hence, $\Delta^x(F_N, G_N)$ can be reasonably proposed as a measure to assess how reliable is the use of a matching distribution for $(Y, Z)|X$ as a surrogate of the actual d.f. That is, as a measure of the reliability of the class (4.27).

From (4.30), $\Delta^x(F_N, G_N)$ can be defined as the maximal error occurring when the true population distribution function $H_N(y, z|x)$ is replaced by a *matching distribution* $\widetilde{H}_N(y, z|x)$ in the class of *plausible* joint p.d.f. (4.27), but with marginal d.f.s estimated on the basis of sample data.

Then the smaller the measure of uncertainty, the more reliable the matching distribution actually used. The most interesting aspect of the uncertainty measure (4.31) is that it can be estimated on the basis of the observed sample data.

A bound similar to (4.30) can be constructed for the unconditional matching error (4.29). In fact, it is immediate to see that

$$ME(H_N^*, H_N) \leq \Delta(F_N, G_N) \tag{4.32}$$

where

$$\Delta(F_N, G_N) = \sum_x \Delta^x(F_N, G_N) p_N(x).$$ (4.33)

is a measure of the *unconditional uncertainty* on the joint population distribution of (X, Y, Z). Clearly, the unconditional measure of uncertainty (4.33) is the average of the conditional measures of uncertainty (4.31), w.r.t. the marginal population distribution of X.

Remark 1. An interesting property of the uncertainty measures (4.31) and (4.33) is that their maximal value can be computed. As it is often the case when dealing with multivariate distribution functions, the use of copulas simplifies matters, as shown, for instance, in [33]. Let H_N be a p.d.f. with margins F_N and G_N, respectively. As already described in [19], there exists a copula C^x such that for all (z, y)

$$H_N(y, z|x) = C^x(F_N(y|x), G_N(z|x)).$$ (4.34)

From Sklar's theorem, if F_N and G_N are continuous, then $C^x(\cdot, \cdot)$ is unique, and it is equal to $H_N\left(F_N^{-1}(y|x), G_N^{-1}(z|x)\right)$. The copula (4.34) represents the "intrinsic" association between Y and Z, disregarding their marginal p.d.f.s. In fact the joint p.d.f. $H_N(y, z|x)$ is given by knowledge of (*i*) the copula (4.34) and of (*ii*) the marginal p.d.f.s $F_N(y|x)$, $G_N(z|x)$. In matching problems, the marginal p.d.f.s $F_N(y|x)$, $G_N(z|x)$ can be estimated on the basis of sample data, the "actual uncertainty" only involves the association expressed by the copula (4.34). The copula version of the Fréchet bounds (4.12) is given by

$$\{C^x(u, v): W^x(u, v) \le C^x(u, v) \le M^x(u, v), \forall u, v \in I\}.$$ (4.35)

Conditionally on x, both $U = F_N(Y|x)$ and $V = G_N(Z|x)$ do have uniform distribution in $(0, 1)$, regardless of the shapes of F_N and G_N. Furthermore, $W^x(u, v) = \max(0, u + v - 1)$, $M^x(u, v) = \min(u, v)$ are themselves copulas; they represent perfect dependence either negative or positive. As a consequence, the maximal value of uncertainty measure (4.31) becomes the volume between the surfaces $M^x(u, v)$ and $W^x(u, v)$:

$$\int_0^1 \int_0^1 \{\min(u, v) - \max(0, u + v - 1)\} \, du \, dv = \frac{1}{6}.$$ (4.36)

The value 1/6 represents the uncertainty achieved when no external auxiliary information beyond knowledge of margins

F_N and G_N is available. The same holds for the unconditional measure of uncertainty (4.33). Roughly speaking, when the only available information is the data, then the uncertainty measure is equal to 1/6, independently of the marginal d.f.s F_N, G_N. This is in accordance with intuition, because on one hand uncertainty *only depends on the maximal and minimal values of the copula* $C^x(u,v)$, *and not on the marginals* F_N, G_N, and on the other hand, *the data do not provide any information on* $C^x(u,v)$. When no extra-sample auxiliary information on $C^x(u,v)$ is available, the minimal and maximal values of $C^x(u,v)$ are $\max(0, u+v-1)$ and $\min(u, v)$, respectively, *independently of the available data*. In a sense, this is the case of *maximal* uncertainty.

Example 4. Contingency tables - Assume that, given a discrete r.v. X with K categories, Y and Z are discrete r.v.s with Ψ and Φ categories, respectively, not necessarily ordered. In this case, the whole theory developed so far can be applied still by simply replacing the d.f.s with probability functions. With no loss of generality, the symbols $k = 1,\ldots,K$, $\psi = 1,\ldots,\Psi$, and $\phi = 1,\ldots,\Phi$, denote the categories taken by X, Y and Z, respectively.

Let $\theta_{\psi\phi|k}$ be the probability that $(Y = \psi, Z = \phi|X = k)$, with marginals $\theta_{\psi.|k}$ representing the probability of the event $(Y = \psi|X = k)$ and $\theta_{.\phi|k}$ representing the probability of the event $(Z = \phi|X = k)$. Since

$$\max(0, \theta_{\psi.|k} + \theta_{.\phi|k} - 1) \leq \theta_{\psi,\phi|k} \leq \min(\theta_{\psi.|k}, \theta_{.\phi|k}) \qquad (4.37)$$

the conditional uncertainty measure turns out to be equal to

$$\Delta^{x=k} = \sum_{\psi=1}^{\Psi}\sum_{\phi=1}^{\Phi}\left\{\min(\theta_{\psi.|k}\,\theta_{.\phi|k}) - \max(0, \theta_{\psi.|k} + \theta_{.\phi|k} - 1)\right\}\theta_{\psi.|k}\theta_{.\phi|k},$$

and the overall uncertainty measure is

$$\Delta = \sum_{k=1}^{K}\Delta^{x=k}\theta_k.$$

where θ_k represents the probability of the event $(X = k)$. Sharper results are obtained when the categories taken by (X, Y, Z) are ordered [17]. For the sake of simplicity, we use the customary order for natural numbers. In this case, the cumulative d.f.s are

$$H_{\psi,\phi|k} = \sum_{\psi=1}^{\Psi}\sum_{\phi=1}^{\Phi}\theta_{\psi\phi|k}, \qquad \psi = 1,\ldots,\Psi, \phi = 1,\ldots,\Phi, k = 1,\ldots,K,$$

$$F_{\psi|k} = \sum_{\psi=1}^{\psi} \theta_{y.|k}, \qquad \psi = 1,\dots,\Psi, \, k = 1,\dots,K,$$

$$G_{\phi|k} = \sum_{\phi=1}^{\phi} \theta_{.z|k}, \qquad \phi = 1,\dots,\Phi, \, k = 1,\dots,K.$$

Then, the following inequalities

$$\max(0, F_{\psi|k} + G_{\phi|k} - 1) \leq H_{\psi,\phi|k} \leq \min(F_{\psi|k}, G_{\phi|k}). \tag{4.38}$$

hold. Note that inequalities (4.38) imply that

$$\theta_{\psi\phi|k}^{-} \leq \theta_{\psi\phi|k} \leq \theta_{\psi\phi|k}^{+}, \tag{4.39}$$

where

$$\theta_{\psi\phi|k}^{-} = L(F_{\psi|k}, G_{\phi|k}) - L(F_{\psi-1|k}, G_{\phi|k}) - L(F_{\psi|k}, G_{\phi-1|k}) + L(F_{\psi-1|k}, G_{\phi-1|k})$$

$$\theta_{\psi\phi|k}^{+} = U(F_{\psi|k}, G_{\phi|k}) - U(F_{\psi-1|k}, G_{\phi|k}) - U(F_{\psi|k}, G_{\phi-1|k}) + U(F_{\psi-1|k}, G_{\phi-1|k}).$$

and given two real numbers a, b, the two quantities L, U are defined as follows

$$L(a,b) = \max(0, a + b - 1), U(a,b) = \min(a,b). \tag{4.40}$$

It is not difficult to prove that

$$\theta_{\psi\phi|k}^{-} \geq L(\theta_{\psi.|k}, \theta_{.\phi|k})$$
$$\theta_{\psi\phi|k}^{+} \leq U(\theta_{\psi.|k}, \theta_{.\phi|k})$$

so that inequalities (4.39) are sharper than (4.37). At any rate, the conditional uncertainty measure is

$$\Delta^{x=k} = \sum_{\psi=1}^{\Psi} \sum_{\phi=1}^{\Phi} \left\{ U(F_{\psi|k}, G_{\phi|k}) - L(F_{\psi|k}, G_{\phi|k}) \right\} \theta_{\psi.|k} \theta_{.\phi|k}, \tag{4.41}$$

and the unconditional uncertainty measure is

$$\Delta = \sum_{k=1}^{K} \Delta^{x=k} \theta_k. \tag{4.42}$$

In contrast with what was found previously, the uncertainty measure is not equal to 1/6, since the uncertainty measure depends on the marginal probabilities of $Y|X$ and $Z|X$. As previously stressed, the same does not hold when Y and Z are continuous. For details, see [17] where the uncertainty in statistical matching for ordered categorical variables is investigated.

4.4   Statistical matching for complex sample surveys

In order to estimate a plausible *matching distribution* as well as to make inference on the uncertainty measures (4.31) and (4.33), it is necessary to make assumptions on the sampling designs according to which the samples s_A, s_B are drawn. Most papers in literature on statistical matching are based on the assumptions that both samples s_A, s_B are composed by *i.i.d.* observations, see for instance [35] and references therein.

Uncertainty in statistical matching in a non-parametric setting for *i.i.d.* observations has been introduced and studied in [16], [17], and [19]. Specifically, in [16] and [19], uncertainty in statistical matching is discussed, and the uncertainty measures (4.31), (4.33) based on Fréchet class are introduced in order to measure the effect on model uncertainty due to the introduction of logical constraints (4.17).

However, the *i.i.d.* assumption is hardly ever valid for survey data, the sample design being frequently complex because of the presence of stratification, different levels of clustering, and different inclusion probabilities based on size measures.

Statistical matching in complex sample surveys is studied in [42] and [39]. The approach proposed by [42] consists in concatenating the two files s_A and s_B by computing new sampling weights relative to an artificial design $p(s)$, where $s = s_A \cup s_B$, given by the union of the sampling designs acting in s_A and s_B, respectively. This approach has been seldom used in practice, since it requires the knowledge of inclusion probabilities of the units in one sample under the sampling design of the other sample. The approach proposed in [39] consists in adapting the actual survey weights of the two distinct samples s_A and s_B in order to have homogeneous common distributions on X, $Y|X$ and $Z|X$ through the repeated application of calibration procedures.

D'Orazio et al. [24] compare their efficiency, including additional estimators based on maximum pseudo-likelihood, as the ones in [48]. Anyway, there is not any prominent estimator among these ones, at finite sample sizes.

When samples are drawn according to complex survey designs, a possibility is given by [18]. They restrict attention to sampling designs of, roughly speaking, high entropy (the actual assumptions are in Section 4.4.1). In this setting, the existence of a plausible joint distribution for the variables of interest can be obtained via iterative proportional fitting (IPF) algorithm (Section 4.4.2). Conti et al. [18] overcome the problem of establishing a best approach for statistical matching for finite sample size by analyzing the reliability of the estimator in terms of asymptotic properties (Section 4.4.3) that allow for a number of additional tools, as the definition of tests for uncertainty width.

4.4.1   Technical assumptions on the sample designs

The assumptions on the sampling designs according to which the samples s_A, s_B are drawn make reference to the following framework.

For each unit i of the population U, let $D_{i,\alpha}$ ($\alpha = A, B$) be a Bernoulli random variable (r.v.), such that i is in the sample s_α if $D_{i,\alpha} = 1$, whilst i is not in s_α if $D_{i,\alpha} = 0$. Let further $\mathbf{D}_{N,\alpha} = (D_{1,\alpha} \dots D_{N,\alpha})$, and ($\alpha = A, B$). A sampling design P_α is the probability distribution of $\mathbf{D}_{N,\alpha}$ ($\alpha = A, B$). In particular $\pi_{i,\alpha} = E_{P_\alpha}[D_{i,\alpha}]$ is the first-order inclusion probability of unit i under the sampling design P_α ($\alpha = A, B$). The (effective) size of s_α is the r.v. $n_{s,\alpha} = D_{1,\alpha} + \cdots + D_{N,\alpha}$ ($\alpha = A, B$). In the sequel, we will only consider fixed (effective) size sampling designs, such that $n_{s,\alpha} \equiv n_\alpha$ ($\alpha = A, B$).

For each unit i, let $p_{i,\alpha}$ be a positive number, with $p_{1,\alpha} + \cdots + p_{N,\alpha} = n_\alpha$ ($\alpha = A, B$). The *Poisson sampling design* Po_α with parameters $p_{1,\alpha}, \dots, p_{N,\alpha}$ is characterized by assuming that the r.v.s $D_{i,\alpha}$s are independent with $Pr_{Po_\alpha}(D_{i,\alpha} = 1) = p_{i,\alpha}$ ($\alpha = A, B$).

The *rejective sampling*, or *normalized conditional Poisson sampling* $P_{R,\alpha}$ (cfr. [27], [47]) is obtained from the Poisson sampling *via* conditioning w.r.t. $n_{s,\alpha} = n_\alpha$ ($\alpha = A, B$). In symbols:

$$Pr_{P_{R,\alpha}}(\mathbf{D}_{N,\alpha}) = Pr_{Po_\alpha}(\mathbf{D}_{N,\alpha} | n_{s,\alpha} = n_\alpha),$$

with $\alpha = A, B$.

The *Hellinger distance* between a sampling design P_α and the rejective design $P_{R,\alpha}$ ($\alpha = A, B$) is defined as

$$d_H(P_\alpha, P_{R,\alpha}) = \sum \left(\sqrt{Pr_{P_\alpha}(\mathbf{D}_{N,\alpha})} - \sqrt{Pr_{P_{R,\alpha}}(\mathbf{D}_{N,\alpha})} \right)^2, \quad \alpha = A, B, \qquad (4.43)$$

where the sum is extended on $D_{1,\alpha}, \dots, D_{N,\alpha}$.

The basic assumptions on sampling designs are the following:

A1. $(\mathcal{U}_N; N \geq 1)$ is a sequence of finite populations of increasing size N.

A2.   For each N, $(x_1, y_1, z_1), \dots, (x_N, y_N, z_N)$ are realizations of a superpopulation $(X_1, Y_1, Z_1), \dots, (X_N, Y_N, Z_N)$ composed by i.i.d. r.v.s (X_i, Y_i, Z_i) with common d.f. $H(x, y, z)$. From now on, the symbol $\mathbb{P}$ is used to denote the (superpopulation) probability distribution of r.v.s (X_i, Y_i, Z_i)s, and $\mathbb{E}$, $\mathbb{V}$ the corresponding operators of mean and variance, respectively. Furthermore, X_i are assumed to be discrete r.v.s, taking values $x^1, \dots, x^K$ with positive probabilities $p(x^1), \dots, p(x^K)$.

A3.   The samples s_A, s_B are independently selected from the population U. Furthermore, s_α is selected according to a fixed size sample design P_α

with positive first-order inclusion probabilities $\pi_{1,\alpha}, \ldots, \pi_{N,\alpha}$ and sample size $n_\alpha = \pi_{1,\alpha} + \cdots + \pi_{N,\alpha}$ ($\alpha = A, B$). Although the sample size, the inclusion probabilities, and the r.v.s $D_{i,A}$s, $D_{i,B}$s, depend on N, for the sake of simplicity in the corresponding symbolism the suffix N will be omitted. It is also assumed that

$$d_{N,\alpha} = \sum_{i=1}^{N} \pi_{i,\alpha}(1 - \pi_{i,\alpha}) \to \infty \text{ and } \frac{1}{N} d_{N,\alpha} \to d_\alpha, \ \alpha = A, B$$

with $0 < d_\alpha < \infty$ ($\alpha = A, B$).

A4. The sample sizes n_A, n_B increase as the population size N does, with

$$\lim_{N \to \infty} \frac{n_\alpha}{N} = f_\alpha \in (0, 1), \ \alpha = A, B.$$

A5. For each population ($\mathcal{U}_N$; $N \geq 1$), let $P_{R,\alpha}$ be the rejective sampling design with inclusion probabilities $\pi_{1,\alpha}, \ldots, \pi_{N,\alpha}$, and let P_α be the actual sampling design, with the same inclusion probabilities as $P_{R,\alpha}$ ($\alpha = A, B$). Then

$$d_H(P_\alpha, P_{R,\alpha}) \to 0 \text{ as } N \to \infty, \alpha = A, B.$$

A6. There exist positive real numbers ς_A ς_B, ζ_A, ζ_B such that

$$\lim_{N \to \infty} \frac{1}{N} \sum_{i=1}^{N} \frac{1}{\pi_{i,\alpha}} = \varsigma_\alpha < \infty, \quad \lim_{N \to \infty} \sum_{i=1}^{N} \frac{1}{(i\pi_{i,\alpha})^2} = \zeta_\alpha < \infty, \ \alpha = A, B.$$

Assumptions A1-A6 are similar to those used in [12] and [13]. They are essentially used to ensure the asymptotic normality of the estimators proposed in the sequel. In particular, assumption A5 implies that the sampling designs P_A, P_B possess maximal asymptotic entropy. Examples of such sampling designs are, as shown in [6] and [5], simple random sampling, successive sampling, Sampford design, Chao design, stratified design, etc. Anyway, as stated in [12] and [13], they can be even too restrictive and used in order to simplify notation and proofs: e.g., neither the independence nor the identical distribution of (X_i, Y_i, Z_i)s are really necessary for their validity.

Given that the main findings in [18] are asymptotic, if the population size N goes to infinity, we must also consider corresponding sequences

$$\mathbf{x}_\infty = (x_1, x_2, \ldots), \ \mathbf{y}_\infty = (y_1, y_2, \ldots), \ \mathbf{z}_\infty = (z_1, z_2, \ldots) \tag{4.44}$$

of x_is, y_is, z_is values, respectively. The actual $\mathbf{x}_N$, $\mathbf{y}_N$, $\mathbf{z}_N$ (4.1) are the "segments" of the first N x_is, y_is, z_is in sequence $\mathbf{x}_\infty$, $\mathbf{y}_\infty$, $\mathbf{z}_\infty$, respectively.

The sequences $\mathbf{x}_\infty$, $\mathbf{y}_\infty$, $\mathbf{z}_\infty$ are produced by the superpopulation model in A2. As a consequence, $\mathbf{x}_\infty$, $\mathbf{y}_\infty$, $\mathbf{z}_\infty$ live in a probability space

$((\mathbb{R}^3)^\infty, \mathcal{B}(\mathbb{R}^3)^\infty, \mathbb{P}^\infty)$, where $\mathcal{B}(\mathbb{R}^3)^\infty$ is the product Borel σ-field over $(\mathbb{R}^3)^\infty$, and $\mathbb{P}^\infty$ is the product measure on $((\mathbb{R}^3)^\infty, \mathcal{B}(\mathbb{R}^3)^\infty)$ generated by $\mathbb{P}$. The probability statements we consider are of the form $Pr_P(\cdot | \mathbf{x}_N, \mathbf{y}_N, \mathbf{z}_N)$, with N going to infinity. The suffix P means that the probability refers to the sampling designs P_A, P_B. The results we will obtain hold for "almost all" sequences $\mathbf{x}_\infty, \mathbf{y}_\infty, \mathbf{z}_\infty$ that the superpopulation model in A2 can produce, i.e., for a set of sequences having $\mathbb{P}^\infty$-probability 1. With a slight lack of precision, we will use the expression "with $\mathbb{P}$-probability 1".

From now on, the symbols

$$F(x, y) = H(x, y, +\infty); \quad G(x, z) = H(x, +\infty, z); \tag{4.45}$$

$$Q(x) = H(x, +\infty, +\infty); \quad p(x) = Q(x) - Q(x^-) \tag{4.46}$$

will denote the joint superpopulation d.f.s of (X_i, Y_i), (X_i, Z_i), the marginal superpopulation d.f. of X_i, and the probability function of X_i, respectively, (note that $p(x) > 0$ iff $x \in \{x^1, \ldots, x^K\}$). Finally, the symbols

$$F(y|x) = \frac{F(x, y)}{p(x)}, \quad G(z|x) = \frac{G(x, z)}{p(x)}, \quad H(y, z|x) = \frac{H(x, y, z)}{p(x)} \tag{4.47}$$

will denote the conditional superpopulation d.f.s of Y_i, Z_i, (Y_i, Z_i) given $X_i = x$, respectively.

4.4.2   A proposal for choosing a matching distribution

The first goal of statistical matching consists in choosing a matching distribution for Y and Z given X, i.e., in the class (4.27) with margins estimated by the samples s_A and s_B, respectively. Under the constraints (4.17), the CIA assumption $H_N(y, z|x) = F_N(y|x)G_N(z|x)$ is not allowed, and it does not seem an easy task to create a distribution fulfilling the constraints.

At a sample level, the d.f.s F_N, G_N are unknown, and must be estimated on the basis of sample data. Conti et al. [18] propose the use of *Hájek estimators*:

$$\widehat{F}_H(y|x) = \frac{\sum_{i=1}^N \frac{D_{i,A}}{\pi_{i,A}} I_{(y_i \le y)} I_{(x_i = x)}}{\sum_{i=1}^N \frac{D_{i,A}}{\pi_{i,A}} I_{(x_i = x)}}, \quad \widehat{G}_H(z|x) = \frac{\sum_{i=1}^N \frac{D_{i,B}}{\pi_{i,B}} I_{(z_i \le z)} I_{(x_i = x)}}{\sum_{i=1}^N \frac{D_{i,B}}{\pi_{i,B}} I_{(x_i = x)}}. \tag{4.48}$$

Next, define the d.f. $\widetilde{H}_H(y, z|x)$ as

$$\int_L d\widetilde{H}_H(y, z|x) = \widehat{C} \int_L I_{(a_x \le f_x(y,z) \le b_x)} d[\widehat{F}_H(y|x)\widehat{G}_H(z|x)] \tag{4.49}$$

where L is any Borel σ-field in $\mathbb{R}^2$ and

$$
\begin{aligned}
\widehat{C}^{-1} &= \int_{\mathbb{R}^2} I_{(a_x \le f_x(y,z) \le b_x)} d[\widehat{F}_H(y|x)\widehat{G}_H(z|x)] \\
&= \frac{\sum_{i \in s_A} \sum_{j \in s_B} I_{(a_x \le f_x(y_i, z_j) \le b_x)} \pi_{i,A}^{-1} \pi_{j,B}^{-1} I_{(x_i = x)} I_{(x_j = x)}}{\sum_{i \in s_A} \sum_{j \in s_B} \pi_{i,A}^{-1} \pi_{j,B}^{-1} I_{(x_i = x)} I_{(x_j = x)}}.
\end{aligned}
$$

In other words, $\widetilde{H}_H(y, z|x)$ can be rewritten as:

$$\widetilde{H}_H(y, z|x) =$$
$$= \frac{\sum_{i \in s_A} \sum_{j \in s_B} I_{(a_x \leq f_x(y_i,z_j) \leq b_x)} \pi_{i,A}^{-1} \pi_{j,B}^{-1} I_{(x_i=x)} I_{(x_j=x)} I_{(y_i \leq y)} I_{(z_j \leq z)}}{\sum_{i \in s_A} \sum_{j \in s_B} I_{(a_x \leq f_x(y_i,z_j) \leq b_x)} \pi_{i,A}^{-1} \pi_{j,B}^{-1} I_{(x_i=x)} I_{(x_j=x)}}$$

The basic idea is to apply the IPF algorithm by taking $\widetilde{H}_H(y, z|x)$ as "starting point", and by alternately re-proportioning its marginals w.r.t. (4.48). Note that the starting distribution $\widetilde{H}_H(y, z|x)$ is a distribution that mimics the conditional independence assumption (due to the choice of differentiating with respect to the product of the distributions $\widehat{F}_H(y|x)$ and $\widehat{G}_H(z|x)$) but does not represent the CIA, due to the assumption that $\widetilde{H}_H(y, z|x)$ is zero outside the constrained region, while the marginal distributions are not necessarily zero in the corresponding domains. Note also that the marginal distributions of $\widetilde{H}_H(y, z|x)$ are not $\widehat{F}_H(y|x)$ and $\widehat{G}_H(z|x)$, respectively. For this reason, the IPF algorithm is applied along $\widehat{F}_H(y|x)$ and $\widehat{G}_H(z|x)$ in order to have as solution an "estimated matching distribution" $\widehat{H}_H^*(y, z|x)$, i.e., a d.f. having at the same time marginals (4.48) and satisfying the constraints (4.17).

Conti et al. [18] propose the use of the estimators (4.48) because, under assumptions A1-A6, they tend in probability to have the same limit as when estimators are based on Poisson sampling design, and in this last case the asymptotic normality of estimators can be easily derived.

4.4.3 Reliability of the matching distribution

In order to assess the accuracy of the matching distribution, we proceed to estimate the uncertainty measures (4.31) and (4.33).

With regard to the conditional measure of uncertainty (4.31), a fairly natural approach consists in estimating first the conditional d.f.s $F_N(y|x)$, $G_N(z|x)$ given by (4.9), and then in plugging such estimates in (4.31). Using the estimators (4.48), the following estimator of the conditional measure of uncertainty $\Delta^x(F_N, G_N)$ is obtained

$$\widehat{\Delta}_H^x = \int_{\mathbb{R}^2} \left(\widehat{K}_{H+}^x(y, z) - \widehat{K}_{H-}^x(y, z) \right) d[\widehat{F}_H(y|x)\widehat{G}_H(z|x)]$$
$$= \frac{1}{\widehat{N}_A(x)\widehat{N}_B(x)} \sum_{i,j=1}^{N} \left(\widehat{K}_{H+}^x(y_i, z_j) - \widehat{K}_{H-}^x(y_i, z_j) \right) \frac{D_{i,A}}{\pi_{i,A}} \frac{D_{i,B}}{\pi_{i,B}} I_{(x_i=x)} I_{(x_j=x)} \quad (4.50)$$

where $\widehat{K}_{H-}^x$, $\widehat{K}_{H+}^x$ are defined exactly as (4.19) and (4.20), respectively, but with F_N, G_N replaced by $\widehat{F}_H, \widehat{G}_H$.

Conti et al. [18] prove Theorem 1, where the large sample distribution of the estimator $\widehat{\Delta}_H^x$ is obtained. Let us define the quantities

$$\widehat{p}_{H,A}(x) = \frac{\sum_{i=1}^{N} \frac{D_{i,A}}{\pi_{i,A}} I_{(x_i = x)}}{\sum_{i=1}^{N} \frac{D_{i,A}}{\pi_{i,A}}}, \quad \widehat{p}_{H,B}(x) = \frac{\sum_{i=1}^{N} \frac{D_{i,B}}{\pi_{i,B}} I_{(x_i = x)}}{\sum_{i=1}^{N} \frac{D_{i,B}}{\pi_{i,B}}}, \tag{4.51}$$

$$\widehat{n}_{H,A}(x) = \frac{N\widehat{p}_{H,A}(x)}{\frac{1}{N}\sum_{i=1}^{N} \pi_{i,A}^{-1} - 1}, \quad \widehat{n}_{H,B}(x) = \frac{N\widehat{p}_{H,B}(x)}{\frac{1}{N}\sum_{i=1}^{N} \pi_{i,B}^{-1} - 1}. \tag{4.52}$$

where $\widehat{p}_{H,A}(x)$ represent the Hájek estimators of $p_N(x)$ obtained from the two samples s_A, s_B, respectively.

Since, as shown in [18]

$$\widehat{n}_{H,A}(x)\left(\frac{n_A p(x)}{f_A(\varsigma_A - 1)}\right)^{-1} \xrightarrow{p} 1, \quad \widehat{n}_{H,B}(x)\left(\frac{n_B p(x)}{f_B(\varsigma_B - 1)}\right)^{-1} \xrightarrow{p} 1 \tag{4.53}$$

as N increases, the asymptotic normality of $\widehat{\Delta}_H^x$ can be established in Theorem 1.

Theorem 1 *Let $x \in \{x^1, \ldots, x^K\}$, and suppose that*

$$\frac{\frac{n_A p(x)}{f_A(\varsigma_A - 1)} \frac{n_B p(x)}{f_B(\varsigma_B - 1)}}{\frac{n_A p(x)}{f_A(\varsigma_A - 1)} + \frac{n_B p(x)}{f_B(\varsigma_B - 1)}} \to \alpha \text{ as } N \to \infty \tag{4.54}$$

Then, for almost all (x_i, y_i, z_i)s values, conditionally on x_N, y_N, z_N, as N increases:

$$\sqrt{\frac{\widehat{n}_{H,A}(x)\widehat{n}_{H,B}(x)}{\widehat{n}_{H,A}(x) + \widehat{n}_{H,B}(x)}} \left(\widehat{\Delta}_H^x - \Delta^x(F_N, G_N)\right) \xrightarrow{d} N(0, V(F, G; x)) \tag{4.55}$$

$N(0, V)$ *denoting a normal r.v. with mean 0 and variance $V(F, G; x)$ given in [18].*

For the unconditional measure of uncertainty (4.33), the idea is to use an estimator of $p_N(x)$ of the form:

$$\widehat{p}_\tau(x) = \tau\widehat{p}_{H,A}(x) + (1 - \tau)\widehat{p}_{H,B}(x) \tag{4.56}$$

with $0 \leq \tau \leq 1$, on the basis of the two observed samples have already been defined in (4.51). A fairly natural choice of τ is the one that minimizes the asymptotic variance of (4.56).

Hence, the proposed estimator of the unconditional measure of uncertainty (4.33) is obtained summing over all the categories x^k $k = 1, \ldots, K$ of X:

$$\widehat{\Delta}_H = \sum_{k=1}^{K} \widehat{\Delta}_H^{x^k} \widehat{p}_{H,AB}(x^k) = \widehat{\mathbf{p}}_{H,AB}(x)'\widehat{\mathbf{\Delta}}_H^x \tag{4.57}$$

that essentially mimics the structure of (4.33).

Theorem 2 *Let*

$$\widehat{n}_A = \frac{n_A}{\frac{n_A}{N}\left(N^{-1}\sum_{i=1}^{N}\pi_{i,A}^{-1}-1\right)} = \frac{N}{N^{-1}\sum_{i=1}^{N}\pi_{i,A}^{-1}-1}, \tag{4.58}$$

$$\widehat{n}_B = \frac{n_A}{\frac{n_B}{N}\left(N^{-1}\sum_{i=1}^{N}\pi_{i,B}^{-1}-1\right)} = \frac{N}{N^{-1}\sum_{i=1}^{N}\pi_{i,B}^{-1}-1}. \tag{4.59}$$

For almost all (x_i, y_i, z_i)s values, conditionally on x_N, y_N, z_N, as N increases the following result holds:

$$\sqrt{\frac{\widehat{n}_A\widehat{n}_B}{\widehat{n}_A+\widehat{n}_B}}\left(\widehat{\Delta}_H - \Delta(F_N, G_N)\right) \xrightarrow{d} N\left(0, V(F, G)\right) \ as \ N \to \infty \tag{4.60}$$

with

$$V(F, G) = \sum_{k=1}^{K} p(x^k)V(F,G;x^k) + \frac{(\varsigma_A-1)(\varsigma_B-1)}{(\varsigma_A+\varsigma_B-2)^2}\Delta^x(F,G)'\Sigma\Delta^x(F,G). \tag{4.61}$$

The estimators $\widehat{\Delta}_H^x$ and $\widehat{\Delta}_H$ are asymptotic design-consistent (in the Brewer sense) as proved in [14].

4.4.4 Evaluation of the matching reliability as a hypothesis problem

The asymptotic results (4.55) and (4.60) allow us to define test hypothesis procedures on the uncertainty measures (4.31) and (4.33), respectively. Given that the conditional uncertainty measure $\Delta^x(F_N, G_N)$ can be interpreted as an upper bound of the error occurring when the true population distribution function is replaced by a matching distribution in $\mathcal{H}_N^x$ (4.27), these test procedures are effective in order to show, for instance, whether the uncertainty can be practically considered as zero, or bounded by a positive negligible number, or if data do not support this hypothesis. Let ϵ^x be such positive number; for instance, it could be a percentage of the maximal value of $\Delta^x(F_N, G_N)$, which is $1/6 \approx 0.17$ (see [16]). A conservative approach would consider as "reliable" a matching distribution whenever $\Delta^x(F_N, G_N)$ is smaller than ϵ^x.

The evaluation of the reliability of a matching distribution can be dealt with in terms of testing the following hypotheses:

$$\begin{cases} H_0: & \Delta^x(F_N, G_N) \le \epsilon^x \\ H_1: & \Delta^x(F_N, G_N) > \epsilon^x \end{cases}. \tag{4.62}$$

Given the (asymptotic) significance level α, the null hypothesis H_0 is accepted if

$$\widehat{\Delta}^x \le \epsilon^x + z_\alpha\sqrt{\widehat{V}_x}\left(\frac{\widehat{n}_{H,A}(x)\widehat{n}_{H,B}(x)}{\widehat{n}_{H,A}(x)+\widehat{n}_{H,B}(x)}\right)^{-1/2}$$

where z_α is the αth quantile of the standard normal distribution, and $\widehat{V}_x$ is an appropriate estimator of the variance $\widehat{V}(F, G; x)$.

Note that the asymptotic variances involved in the previous formulas are awkward. A proposal is to construct the estimator $\widehat{V}_x$ either by the usual plug-in rule (replace $F_N(y|x)$ and $G_N(z|x)$ in variance expression by $\widehat{F}_H(y|x)$ and $\widehat{G}_H(z|x)$, respectively), or by resampling methods for complex designs (cfr. [3], [13]). Clearly, similar considerations hold for the unconditional uncertainty measure (4.33), using (4.60).

Remark 2. Under the assumption *i.i.d.* the classical version of bootstrap can be applied to estimate the variance, as applied in [18]. It consists in (i) generate a sample of size n_A form s_A; (ii) generate a sample of size n_B form s_B; use such samples to compute the bootstrap version of uncertainty measure can be applied. Clearly, the classical bootstrap method, as proposed by [25], does not work in complex survey sampling. Conti et al. [18] detail also the references for resampling approaches from finite populations, mentioning two main approaches: the *ad hoc* approach ([31], [37], [44], [4], [11], [13]) and the *plug in* approach ([26], [10], [28], [36]).

4.5 Conclusions and pending issues: relationship between the statistical matching problem and ecological inference

Statistical matching is a problem that is going to become ubiquitous as long as different data sources start to share a common metadata framework, and in a unique place it is possible to see data that neither has been jointly observed nor can be linked by unit identifiers. Although first applications appeared in the late 1960s, statistical matching applications are not very common due to important problems: data cannot provide point estimates for parameters that describe the joint relationship of the never jointly observed variables, unless specific assumptions are introduced; data can be possibly obtained by means of different complex survey designs and it is not straightforward to reconcile them.

As already introduced in Section 4.2, statistical matching tackles the problem of inferring the joint distribution of two variables never jointly observed. This problem has been independently analyzed in a different framework: ecological inference. There are some aspects that characterize ecological inference that make this framework specific: anyway, the contact points between the two areas are so many that it seems to be important to analyze them together and possibly exchange methods and solutions. Ecological inference "draws conclusions concerning individual-level relationships using data in the form of aggregates for groups in the population. The groups are often geographically defined" [46]. The term *ecological* seems to be due to the presence of the groups, and the fact that they are generally geographically designed. As a matter of fact, statistical matching does not refer to any kind of grouping, but uses the common variables X for corroborating any kind of inference on the variables never jointly observed. Despite these similarities, solutions have not been shared in the two contexts. This is also represented by the typical ecological inference example: electoral votes for two parties are known at the time of the elections, as well as the number of voters according to ethnicity (whites, non-whites), for a number of precincts. Consequently, two-way contingency tables for ethnicity and electoral votes can be built, where the margins are known and the cells inside the tables are unknown. This is exactly the statistical matching situation with X representing precincts. As already pointed out in the statistical matching context, the joint distribution of ethnicity and votes given precincts, i.e., $(Y, Z|X)$ according the statistical matching notation, is the unidentifiable distribution. That is the object of interest also for ecological inference.

Percentages inside the contingency tables given precincts, when nothing is known, are values between zero and one. When marginals are known for each precinct, the space of admissible values for the percentages inside the contingency tables reduce to a line, that King [29] calls tomography line. This corresponds to the reduction of uncertainty in statistical matching for pairs of binomial random variables given their conditional margins.

A slight difference is that the focus of ecological inference is on already aggregated data at the group (precinct) level. Hence, King [29] assumes a model that generates conditional percentages inside each precinct and works on that. On the contrary, statistical matching uses to model directly any single observation in any of the tables ([46] formalize this approach, distinguishing between sampling at individual and aggregate level). Furthermore, the groups X seem to be given, while in statistical matching they are modeled together, the variables of interest and sometimes the object of interest in statistical matching is just on the pairwise distribution (Y, Z). Note that the notion of uncertainty about the joint distribution of Y and Z is strictly related to the association between X with Y and Z. Possibly, this slight change of perspective produced independent solutions in the two areas, therefore it is time to conduct both areas in the same framework.

For instance, in ecological inference, King [29] makes assumptions on the percentages inside tables for each precinct that can be summarized in this way:

1. the two conditional percentages of $Y|Z$ for each precinct jointly follow a specific distribution (e.g., the truncated normal)

2. conditional on the observed percentage on ethnicity, the observed percentage on voters in different precincts are mean independent (no spatial autocorrelation)

3. the observed marginal proportions on ethnicity are independent of the two conditional percentages modeled in point 1.

In this setting, it is possible to perform maximum likelihood or Bayesian analyses on the conditional probabilities of voters given ethnicity. Statistical matching, on the contrary, has been developed as much as possible as an assumption-free framework (or better, the initial assumptions on the conditional independence of Y and Z given X were clearly unsatisfactory). It is time that also the statistical matching framework can be built on reasonable assumptions. As in ecological inference or other areas (e.g., small area models) when data is not able to support the target of the analyses, assumptions are not necessarily evil, even in the mostly assumption-free official statistics framework. Hence, methods developed in ecological inference should begin to be applied also for statistical matching purposes. Some methodological challenges remain, and this is just a preliminary list:

- the approach derived from ecological inference suggests that it is possible to rethink the role of the common variables in the two sample surveys to match: from the common variables it is possible to consider covariates (i.e., matching variables) on the one hand, and groups (strata) in the second. The second group is fundamental for the possibility of modeling aggregate data;

- it seems that ecological inference has been developed for just binary Y and Z, while statistical matching does not restrict the nature of the variables to match.

In our opinion, these are the lines along which the exchange and reuse of methods in the two areas should go.

Bibliography

[1] A. Agresti and M.C. Yang. An empirical investigation of some effects of sparseness in contingency tables. *Computational Statistics & Data Analysis*, 5:9–21, 1987.

[2] T.W. Anderson. Maximum likelihood estimates for a multivariate normal distribution when some observations are missing. *Journal of the American Statistical Association*, 52:200–203, 1957.

[3] E. Antal and Y. Tillé. A direct bootstrap method for complex sampling designs from a finite population. *Journal of the American Statistical Association*, 106:534–543, 2011.

[4] J-F. Beaumont and Z. Patak. On the Generalized Bootstrap for Sample Surveys with Special Attention to Poisson Sampling. *International Statistical Review*, 80:127–148, 2012.

[5] Y.G. Berger. Asymptotic consistency under large entropy sampling designs with unequal probabilities. *Pakistan Journal of Statistics*, 27:407–426, 2011.

[6] Y.G. Berger. Rate of convergence to normal distribution for the Horvitz-Thompson estimator. *Journal of Statistical Planning and Inference*, 67:209–226, 1998.

[7] W. Bergsma. A bias-correction for Cramer's V and Tschuprow's. *Journal of the Korean Statistical Society*, 42:9–21, 2013.

[8] L. Breiman. Random forests. *Machine Learning*, 45:5–32, 2001.

[9] L. Breiman, J. H. Friedman, R. A. Olshen, and C. J. Stone. *Classification and Regression Trees*. Wadsworth, 1984.

[10] M.-T. Chao and S.-H. Lo. A bootstrap method for finite population. *Sankhyā*, pages 399–405, 1985.

[11] A. Chatterjee. Asymptotic properties of sample quantiles from a finite population. *Annals of the Institute of Statistical Mathematics*, 63:157–179, 2011.

[12] P.L. Conti. On the estimation of the distribution function of a finite population under high entropy sampling designs, with applications. *Sankhyā* B. DOI: 10.1007/s13571-014-0083-x, 2014.

[13] P.L. Conti and D. Marella. Inference for quantiles of a finite population: asymptotic vs. resampling results. *Sandinavian Journal of Statistics*, 2014.

[14] P.L. Conti, D. Marella, and A. Neri. Statistical matching and uncertainty analysis in combining household income and expenditure data. *Statistical Methods & Applications*, 26:485–505, 2017.

[15] P.L. Conti, D. Marella, and M. Scanu. Evaluation of matching noise for imputation techniques based on nonparametric local linear regression estimators. *Computational Statistics and Data Analysis*, 53:354–365, 2008.

[16] P.L. Conti, D. Marella, and M. Scanu. Uncertainty analysis in statistical matching. *Journal of Official Statistics*, 28:69–88, 2012.

[17] P.L. Conti, D. Marella, and M. Scanu. Uncertainty analysis for statistical matching of ordered categorical variables. *Computational Statistics and Data Analysis*, 68:311–325, 2013.

[18] P.L. Conti, D. Marella, and M. Scanu. Statistical matching analysis for complex survey data with applications. *Journal of the American Statistical Association*, 111:1715–1725, 2016.

[19] P.L. Conti, D. Marella, and M. Scanu. How far from identifiability? a systematic overview of the statistical matching problem in a non parametric framework. *Commmunications in Statistics - Theory and Methods*, 46:967–994, 2017.

[20] T. De Waal. Statistical matching: experimental results and future research questions. *Discussion paper n. 19, CBS*, 2015.

[21] G. Donatiello, M. D'Orazio, D. Frattarola, A. Rizzi, Scanu M., and M. Spaziani. The role of the conditional independence assumption in statistically matching income and consumption. *Statistical Journal of the IAOS*, 77:667–675, 2016.

[22] M. D'Orazio, M. Di Zio, and M. Scanu. Auxiliary variable selection in a statistical matching problem. In L.-C. Zhang and R. L. Chambers, editors, *Ecological Inference: new methodological strategies*. CRC: Chapman and Hall, London.

[23] M. D'Orazio, M. Di Zio, and M. Scanu. *Statistical matching: theory and practice*. Wiley, Chichester, 2006.

[24] M. D'Orazio, M. Di Zio, and M. Scanu. Uncertainty intervals for nonidentifiable parameters in statistical matching. *Proceedings of the 57th Session of the International Statistical Institute World Congress, Durban - South Africa*, 2009.

[25] B. Efron. Bootstrap methods: another look at the jackknife. *The Annals of Statistics*, 7:1–26, 1979.

[26] S.T. Gross. Median estimation in sample surveys in: Proceedings of the section on survey reasearch methods. *Proceedings of the Section on Survey Research Methods, American Statistical Association*, pages 181–184, 1980.

[27] J. Hájek. Asymptotic theory of rejective sampling with varying probabilities from a finite population. *The Annals of Mathematical Statistics*, 35:1491–1523, 1964.

[28] A. Holmberg. A bootstrap approach to probability proportional-to-size sampling. *Proceedings of the ASA Section on Survey Research Methods*, pages 378–383, 1998.

[29] G. King. *A solution to the ecological inference problem: reconstructing individual behavior from aggregate data*. Princeton University Press, Princeton, 1997.

[30] D. Marella, M. Scanu, and P.L. Conti. On the matching noise of some nonparametric imputation procedures. *Statistics and Probability Letters*, 78:1593–1600, 2008.

[31] P J. McCarthy and C B. Snowden. The bootstrap and finite population sampling. In *Vital and health statistics*, pages 1–23. Public Heath Service Publication, U.S. Government Pronting, Washington, DC, 1985.

[32] C. Moriarity and F. Scheuren. Statistical matching: A paradigm for assessing the uncertainty in the procedure. *Journal of Official Statistics*, pages 407–422, 2001.

[33] R.B. Nelsen. *An introduction to copulas*. Springer, New York, 1999.

[34] B.A. Okner. Constructing a new data base from existing microdata sets: the 1966 merge file. *Annals of Economic and Social Measurement*, 1:325–342, 1972.

[35] S. Raessler. *Statistical matching: A Frequentist Theory, Practical Applications, and Alternative Bayesian Approaches*. Springer, New York, 2002.

[36] M. G. Ranalli and F. Mecatti. Comparing recent approaches for bootstrapping sample survey data: a first step towards a unified approach. In *Proceedings of the ASA Section on Survey Research Methods*, pages 4088–4099, 2012.

[37] J N K. Rao and C F J. Wu. Resampling inference with complex survey data. *Journal of the American Statistical Association*, 83:231–241, 1988.

[38] J. Reiter. Using multiple imputation to integrate and disseminate confidential microdata. *International Statistical Review*, 77:179–195, 2009.

[39] R.H. Renssen. Use of statistical matching techniques in calibration estimation. *Survey Methodology*, 24:171–183, 1998.

[40] D. Rivers. Sampling fom web surveys. In *Proceedings of the ASA Section on Survey Research Methods*, 2007.

[41] W. L. Rodgers. An evaluation of statistical matching. *Journal of Business and Economic Statistics*, pages 91–102, 1984.

[42] D B. Rubin. Statistical matching using file concatenation with adjusted weights and multiple imputations. *Journal of Business and Economic Statistics*, 4:87–94, 1986.

[43] C.A. Sims. Comments on: "Constructing a new data base from existing microdata sets: the 1966 merge file", by B.A. Okner. *Annals of Economic and Social Measurements*, 1:343–345, 1972.

[44] R R. Sitter. A resampling procedure for complex data. *Journal of the American Statistical Association*, 87:755–765, 1992.

[45] D. Slepian. The one-sided barrier problem for gaussian noise. *Bell System Technical Journal*, 41:463–501, 1962.

[46] D.G. Steel, E.J. Beh, and R. L. Chambers. The information in aggregate data. In G. King, O. Rosen, and M.A. Tanner, editors, *Ecological Inference: new methodological strategies*. Cambridge University Press, Cambridge, 2004.

[47] Y. Tillé. *Sampling Algorithms*. Springer, New York, 2006.

[48] C. Wu. Combining information from multiple surveys through the empirical likelihood method. *Canadian Journal of Statistics*, 32:15–26, 2004.

[49] L.-C. Zhang and R. L. Chambers. Minimal inference from incomplete 2 × 2-tables. In L.-C. Zhang and R. L. Chambers, editors, *Analysis of Integrated Data*. CRC: Chapman and Hall, London.

5

Auxiliary variable selection in a statistical matching problem

Marcello D'Orazio

Istat, National Institute of Statistics, Rome, Italy

Marco Di Zio

Istat, National Institute of Statistics, Rome, Italy

Mauro Scanu

Istat, National Institute of Statistics, Rome, Italy

CONTENTS

5.1 Introduction

Statistical matching (SM, sometimes called *data fusion* or *synthetical matching*) aims at combining information available in distinct sample surveys referred to the same target population when the two samples are disjoint. Formally, let Y and Z be two random variables; statistical matching techniques have the objective to estimate the joint (Y, Z) probability distribution function or one of its parameters (e.g., a contingency table or a regression coefficient) when:

(i) Y and Z are not jointly observed in a survey, but Y is observed in a sample A, of size n_A, and Z is observed in a sample B, of size n_B;

(ii) *A* and *B* are independent and units in the two samples do not overlap (it is not possible to use record linkage);

(iii) *A* and *B* both observe a set of additional variables *X*.

This problem was first studied methodologically in [3] in the case of a trivariate (X, Y, Z) Gaussian distribution. Studies on the lack of identifiability of this problem date back to [24]. Lack of identifiability has the effect that multiple equally plausible estimates of the joint (Y, Z) distribution are available. Traditional approaches either implicitly or explicitly introduce assumptions to make the model identifiable in order to obtain a unique estimate of the model parameters. The usual assumption is the conditional independence of *Y* and *Z* given *X*. However, this is a strong assumption that – given the data at hand – cannot be tested. In order to avoid the introduction of assumptions with the consequence of making the inferences more credible, a number of papers have started studying how to make inference taking into account the non-uniqueness of estimates in statistical matching (see also [12]). These inferences are about finding out which values of the true parameter of interest are compatible with the observations we made. This set of values is either named "uncertainty region" [15] or "partial identification region" [31, 17]. Inference based on partial identification regions is used in other contexts such as for instance econometrics [30] and social sciences [20]. It is worthwhile to remark that the "uncertainty regions" should not be confused with the regions determined by "confidence intervals". In the first case, they arise from partial identifiability of the joint distribution; in the second case, the intervals are determined by the sampling nature of the observations. Results on a joint analysis of these two types of uncertainty are described in [19] and [28].

In statistical matching, Rubin [25] defines a non-proper Bayesian approach for the exploration of uncertainty regions. This method was generalized in [23] by presenting proper Bayesian approaches. Moriarty and Scheuren [21] explore the set of equally plausible solutions fixing all the estimable parameters equal to the estimates obtained by means of consistent estimators; this approach is not completely satisfactory because some of the results are not admissible (e.g., covariance matrices in case of Gaussian variables that are negative-definite). The use of maximum likelihood estimators for estimating the identifiable parameters and then finding the corresponding likelihood ridge (i.e., the set of equally maximum likelihood estimates) avoids the possibility to include non-admissible solutions in this set (see [16]). The use of the term *uncertainty* in order to describe the width of the likelihood ridge was first proposed in [16]. Its properties in different contexts and its estimation together with the study of the asymptotic properties of the estimators are in [11]. Alternative approaches related to the concept of minimal inference are outlined in [33]. Although uncertainty is a notion useful to describe how far from the case of identifiability we are in the statistical matching context, the idea of this chapter is that it can also be used for operational purposes, that is for variable selection in SM. More precisely, assume that the two surveys A and B observe a sufficiently large number of common variables X. The case of a large number of common variables X is quite frequent: e.g., in case of social surveys many socio-economic variables (residence, age, gender, professional, educational and marital status, characteristics of the head of the household, characteristics of the households and so on) are available. Should all these variables be used for matching purposes? Problems may arise; for instance, it is known that the larger the number of categorical variables, the higher is the risk of having sparse tables, i.e., tables with cells with few or zero observations per cell. More in general, a high number of variables and consequently of categories may have impact on the efficiency of estimates. In this case, the main problem consists in estimating simultaneously the high number of identifiable parameters of (Y,Z) given X. Classic methods for variable selection are based on the analysis of the explicative power of auxiliary variables X in terms of (Y,Z), for instance by analysing the residuals of the model. Nevertheless, classic methods require joint observations of (Y,Z) that are not available in the case of statistical matching. The consequence is that it is usually adopted as a suboptimal approach; that is, to select variables X explaining Y and Z separately, in fact available data allow to apply classic methods in these two separate datasets. In this chapter we claim that an appropriate selection of the matching variables can be found through the notion of uncertainty, that is to select those variables that minimize the uncertainty region. In order to avoid selecting all the available common variables, a penalised measure of uncertainty should be taken into account. In this chapter, we study penalisations based on sparseness of tables.

5.2 Choice of the matching variables

Data sources A and B may share many common variables X. In performing SM, not all the X variables will be used but just the most important ones. The selection of the most relevant X_M ($X_M \subseteq X$), called *matching variables*, is usually performed by consulting subject matter experts and through appropriate statistical methods (see [14]).

The choice of the matching variables should be made in a multivariate sense [8] to identify the subset X_M connected, at the same time, with Y and Z. This would require the availability of a data source in which (X,Y,Z) are observed. Unfortunately, in the basic SM framework this is not possible because (X,Y,Z) are never jointly observed, and consequently the selection is made by selecting variable in the two datasets A and B separately. Our proposal is to perform a unique analysis for choosing the matching variables by searching the set of common variables that are the most effective in reducing the *uncertainty* between Y and Z, avoiding selecting too many X variables which results in a large number of parameters to estimate.

5.2.1   Traditional methods based on association

Variable selection methods are generally based on association/dependence analysis, and analysis of the residuals of dependent variables. As aforementioned, in the basic SM framework, this is not allowed because of the lack of joint observations of (Y, Z), A only permits to investigate the relationship between Y and X, while the relationship between Z and X can be studied in B. Given these premises, two separate analyses are carried out on A and B to select the corresponding X variables, and the results of the two separate analyses are then joined and, in general, the following rule can be applied:

$$X_Y \cap X_Z \subseteq X_M \subseteq X_Y \cup X_Z$$

where X_Y ($X_Y \subseteq X$) and X_Z ($X_Z \subseteq X$) are the subsets of the common variables that better explain Y and Z, respectively. The intersection $X_Y \cap X_Z$ provides a smaller subset of matching variables if compared to $X_Y \cup X_Z$; this is an important feature in achieving parsimony. For instance, too many matching variables in a distance hot-deck SM micro-application can introduce undesired additional noise in the final results. Unfortunately, the risk with $X_Y \cap X_Z$ is that the best predictors of one target variable will be excluded if they are not in the subset of the predictors of the other target variable, moreover, in some cases, the intersection may be empty. For this reason, the final subset of the matching variables X_M is usually a compromise and the contribution of subject matter experts and data analysts is important in order to identify the "best" subset.

The simplest procedure to identify X_Y consists in calculation of pairwise correlation/association measures between Y and each of the available predictors X. The same analysis should be performed on B to identify the best predictors of Z. When the response variable is continuous, one can look at correlation with the predictors. In order to identify eventual nonlinear relationship, it may be convenient to consider the ranks (Spearman's rank correlation coefficient). An interesting suggestion – that can be found in [18] – consists in looking at the adjusted R^2 related to the regression model rank(Y) vs. rank(X) (unadjusted R^2 corresponds to squared Spearman's rank correlation coefficient). When X is a categorical nominal variable, it is considered the adjusted R^2 of the regression model rank(Y) vs. dummies(X).

When response and predictors are all categorical, then Chi-square based association measures (Cramer's V) or measures of *proportional reduction of the variance* (see e.g., Goodman–Kruskal λ or τ) can be considered. Sakamoto and Akaike [26] suggest using Akaike information criterion (AIC), i.e., the best predictor is the one giving the minimum of the AIC statistic. The nice feature in the AIC-based selection procedure is that it permits also identifying the optimal combination of predictors.

Sometimes, the important predictors can be identified by fitting models and then running procedures for selecting the best predictors. The selection of the subset X_Y can also be demanded to non-parametric procedures such as *Classification and Regression Trees* [6] or, better, *random forest* [7] which provides a measure of importance for the predictors. Fitting a random forest is, however, computationally demanding and some authors warn about using predictor importance measures in selecting the best predictors (see, e.g., [29]).

5.2.2 Choosing the matching variables by uncertainty reduction

In SM there is an intrinsic uncertainty due to the structure of the datasets at hand: estimators of the parameters describing the association/correlation between Y and Z report multiple equally plausible solutions. For instance, estimates obtained by a maximum likelihood estimator of these parameters are in a usually closed set of equally maximum likelihood solutions known as likelihood ridge. The non-uniqueness of the solution of the SM problem has been described in different articles (see [15] and Chapter 4 in [16]).

Given that A and B do not contain any information on Y and Z, apart from their association/correlation with the common variables X, the set of solutions describes all the values of the parameters represented by all the possible relationships between Y and Z given the observed data. For this reason, D'Orazio et al. [16] called this set of equally plausible estimates as "the uncertainty set". The idea we follow for variable selection is that of finding the set of variables X that minimize the uncertainty of statistical matching problem, that is the set of X leads to the smallest uncertainty set.

When X, Y and Z are categorical, the uncertainty set can be computed by resorting to the Fréchet bounds. Let $p_{hjk} = Pr(X = h, Y = j, Z = k)$ for $h = 1,\ldots,H$, $j = 1,\ldots,J$, $k = 1,\ldots,K$; by conditioning on X, the probability $p_{.jk} = Pr(Y = j, Z = k)$ lies in the interval:

$$[\underline{p}_{.jk}, \overline{p}_{.jk}] = \left[\sum_h p_{h..} \max\{0, p_{j|h} + p_{k|h} - 1\}, \sum_h p_{h..} \min\{p_{j|h}, p_{k|h}\} \right] \quad (5.1)$$

It should be noted that when external information is available, for instance in terms of structural zeros on cells of the contingency table $Y \times Z$ or $Y \times Z | X$, these bounds are not sharp; in fact, the admissible values are a closed sub-interval (when the estimated probabilities are compatible), see [31].

The expression (5.1) allows to derive bounds for each cell in the contingency table $Y \times Z$. This information can be used to derive an overall measure of uncertainty; a very basic one can be obtained by considering the average of the bounds' width:

$$d = \frac{1}{J \times K} \sum_{j,k} \left(\overline{p}_{.jk} - \underline{p}_{.jk} \right) \quad (5.2)$$

This is a simple and straightforward way of measuring uncertainty, but it is not unique. In fact, it suffers from the fact that each bound is computed cell by cell, without taking into account what happens to the other cells of the contingency table. A general uncertainty measure for the whole distribution of (Y, Z) given X, and not just for a cell of this distribution, valid in general whatever the nature of the variables at hand, i.e., discrete or continuous, has been proposed in [9]. When Y and Z are continuous, not necessarily Gaussian, non-parametric estimators are proposed, as discussed in [11]. The possibility to apply logical constraints in these settings is illustrated. Estimators in case of samples drawn according to complex survey designs and their asymptotic properties are then illustrated in [10]. Other measures of uncertainty are in [32].

The method proposed for selecting matching variables when dealing with categorical X, Y and Z variables relies on a simple idea: select the subset of the X variables that is more effective in reducing the uncertainty measured, for instance in terms of d. We aim at identifying the best set of X minimising the set of competing models. Unfortunately, the value of d decreases by increasing the number of X variables, and in principle we could select the whole set of X, but since we need to perform an estimation step, it is important to refer to the parsimony principle. In fact, although inference under models with too few parameters (variables) can be biased, models having too many parameters may suffer poor precision or identifying effects that are, in fact, spurious. For this reason, it is necessary to identify a criterion to decide how to restrict the selection of variables X.

Under a different point of view, imagine a case where $Y|X$ and $Z|X$ are perfectly known (which can be expected asymptotically). In this extreme case, the only attention should be devoted to the width of the uncertainty set, because the estimable distributions that determine the Fréchet bounds are without error. When just the two samples A and B are known, the additional aspect of the accuracy of the estimates of the estimable distributions $Y|X$, $Z|X$ and X and consequently of the uncertainty d becomes an important issue to take into consideration. We take care of this aspect in the penalty term, for instance by ensuring that the sparseness is below a certain level, and consequently the estimators still behave quite well.

5.2.3   An illustrative example

An illustration of the problem is given by the results of this empirical analysis based on simulations. The reference data are those from "Quine" [22] consisting of 146 observations and 5 variables. Variable "Eth" (*ethnic background*) has two categories: *Aboriginal or Not*; "Sex" with two categories, "Age" group with 4 categories, "Lrn" (*learner status*) is a factor with 2 levels (*Average* or *Slow learner*), and "Days" that is the number of days absent from school in the year. The latter variable is grouped in 7 categories and named ("c.Days").

In order to transform data for a statistical matching problem, the common variables X are chosen to be "Eth", "Sex" and "Age", while "Lrn" is Y and Z is "c.Days".

We refer to the whole set of data as the target population, and consequently the uncertainty measure is computed by means of the relative frequencies of (Y, X) and (Z, X) observed on the 146 units. We denote this quantity with d^*.

Then we start with the statistical matching exercise, and we sample without replacement two set of data (A and B). The uncertainty measure d is estimated by using A and B, we denote this estimate with $\hat{d}$. The quantities d^* and $\hat{d}$ are computed according to the different combination of variables X.

This process is replicated 10000 times for two different settings $S1$ and $S2$ that are distinct by the sample size of A and B; more precisely in $S1$ the sample size of A and B are $n_A = 30$ and $n_B = 20$, while in $S2$ they are $n_A = 70$ and $n_B = 79$.

Figures 5.1 and 5.2 show the boxplots of the $\hat{d}$ and the target value d^* that is reported with the "*" symbol.

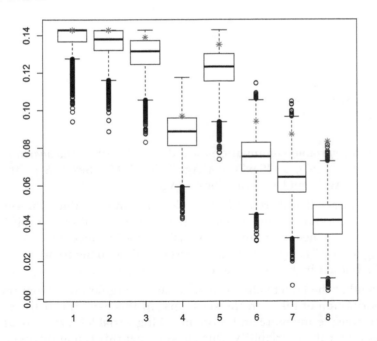

FIGURE 5.1
Simulated example: n_A=70, n_B=79.

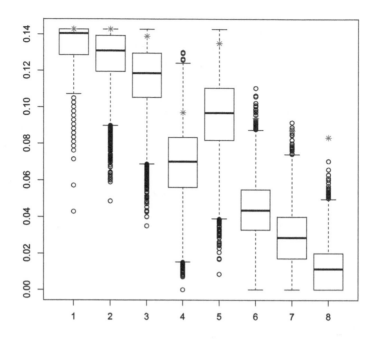

FIGURE 5.2
Simulated example: $n_A=30$, $n_B=20$.

The matching variable combinations are denoted with 1 (*no matching variables*), 2 (only "Eth"), 3 ("Sex"), 4 ("Age"), 5 ("Eth", "Sex"), 6 ("Eth", "Age"), 7 ("Sex", "Age"), and 8 ("Eth", "Sex", "Age").

Firstly, we observe that there is always a decrease in the value of uncertainty as the number of matching variables increases. We remind that this is true only for the sets including the same variables, for instance uncertainty corresponding to "Age" is higher than that corresponding to the combination ("Eth", "Age"), but not higher than that of ("Eth", "Sex").

We observe that the target value d^* is more and more far away from the empirical distribution of the $\hat{d}$, in the sense that in this experiment the estimates, when considering more and more matching variables, are more affected by bias rather than variability. Our guess is that this is mainly due to the presence of sampling zero cells corresponding in fact in the population to non-zero cells. This would explain the systematic underestimation observed in the box-plot. Comparing the two figures, we observe that the performance of $\hat{d}$ gets worse with the decrease of the sample sizes.

According to the previous considerations, we propose to add a penalty term to d related to the accuracy of the estimators, and in particular to the sparseness of the contingency tables obtained by the matching variables.

5.2.4 The penalised uncertainty measure

Before illustrating the procedure, let us introduce some notations. Let Q be the number of all the available X variables and H_q the number of categories of the variable X_q. Moreover, let D_m be the variable obtained by cross-classifying a subset of common variables being H_{D_m} the number of its categories; the total number of subsets that can be created is $M = (2^Q - 1)$, whereas the final subset includes all the Q variables, i.e., $D_Q = X_1 \times X_2 \times \ldots \times X_Q$, and has H_{D_Q} categories ($H_{D_Q} = H_1 \times H_2 \times \ldots \times H_Q$).

The matching variables are those corresponding to the subset D_m such that a penalised measure of d is minimised over all the possible combinations of variables. More formally, let g be a penalty term, we search for the minimum of

$$\tilde{d}_m = d_m + g \tag{5.3}$$

In particular, we have analysed two alternative penalised measures, both of them are based on a measure of sparseness. The first is

$$\tilde{d}_m^{(1)} = d_m + log\left(1 + \frac{H_{D_m}}{H_{D_Q}}\right) \tag{5.4}$$

and we search for the minimum of $\tilde{d}_m^{(1)}$ having an acceptable degree of sparseness.

For instance, if sparseness is measured by

$$\bar{n} = min\left[\frac{n_A}{H_{D_m} \times J}, \frac{n_B}{H_{D_m} \times K}\right] \tag{5.5}$$

where $H_{D_m} \times J$ and $H_{D_m} \times K$ are the number of cells in the table $D_m \times Y$ and $D_m \times Z$, respectively, then we may require that the acceptable subsets are those with $\bar{n} > 1$.

In this function, sparseness is considered in two parts, in the penalty term $log\left(1 + \frac{H_{D_m}}{H_{D_Q}}\right)$, where the sparseness is about variables X and in the stopping rule requiring $\bar{n} > 1$, that is the sparseness related to (Y, X) and (Z, X). The logarithm is introduced because from the experiments performed the rate of increase of this term (without taking logarithm) was noted to be too fast with respect to d_m.

The second penalised measure is

$$\tilde{d}_m^{(2)} = d_m + max\left[\frac{1}{(n_A - H_{D_m} \times J)}, \frac{1}{(n_B - H_{D_m} \times K)}\right] \tag{5.6}$$

with $n_A > (H_{D_m} \times J)$ and $n_B > (H_{D_m} \times K)$.

The second penalization term has the advantage of not being dependent on the number of cells (H_{D_Q}) of the largest table, obtained when crossing all the available X variables. In addition, the second penalty implicitly includes this stopping criterion, i.e., stops when $\bar{n} \leq 1$, representing an indication of a too sparse contingency table. This is certainly a positive aspect of $\tilde{d}_m^{(2)}$, on the other hand, $\tilde{d}_m^{(1)}$ is more flexible since it allows to adopt a different measure of sparseness and different thresholds in the stopping rule. In fact, the criterion $\bar{n} > 1$ is a subjective choice which reflects the broad definition of sparseness in [2]: "contingency tables are said to be sparse when the ratio of the sample size to the number of cells is relatively small". In literature, however, many alternative methods have been proposed to measure sparseness, e.g., measures based on the percentage of expected cell frequencies smaller than 1, 5 or 10, or the percentage of observed zero frequencies. Those indicators could represent alternative proposals for the penalisation term.

The computation of the minimum of the penalised measure may be unfeasible when many common variables characterise the problem; hence, a sequential procedure to approximate the previous computation is proposed. It consists of the following steps:

- **Step 0**

  Sort the Q variables X_q for $q = 1, \ldots, Q$ in descending order with respect to $\tilde{d}_q$

  Then the variable with the smallest $\tilde{d}_q$ is selected.

- **Step 1**

  Consider all the possible combinations obtained by adding one more variable to the current selected set of variables and evaluate the corresponding uncertainty in terms of $\tilde{d}_m$; e.g., in the first iteration all the possible combinations of the variable identified in Step 0 with the others will be considered.

- **Step 2**

  Select the combination of variables with the minimum value of $\tilde{d}_m$, then stop if one of the following conditions is verified:

  - i) $\tilde{d}_m$ is greater than the one corresponding to the combination of the previous iteration, i.e., there are not any combinations of variables returning a decrease of $\tilde{d}_m$,

  - ii) (*step required only in case $\tilde{d}_m^{(1)}$ is used*) the degree of sparseness is not acceptable, for instance by using formula (5.5), $\bar{n} \leq 1$;

  otherwise go back to Step 1.

5.3   Simulations with European Social Survey data

In order to show how the methods for the selection of matching variables work, a toy example with data extracted from round 7 of the European Social Survey (ESS, 2014) is considered. The dataset includes just a subset of the variables in the socio-demographic core module and missing values, "Don't know", refusals have been discarded. In summary, there are $n = 28,769$ respondents and 13 variables (see Table 5.1).

The latter two variables in Table 5.1 play the role of target ones (Y="hincfel1" and Z="hinctnta"), while the others are treated as available common variables (X). Initially, we will use the whole sample for the selection of the matching variables, a condition that is not encountered in real SM application where there are only two independent samples A and B, including (X, Y) and (X, Z), respectively.

At a first step, we look for the ideal set of best predictors of the target variable obtained by crossing the two target ones (i.e., $W = Y \times Z$) by computing the usual pairwise measures. This results in the target covariates we would like to find as predictors, since it is based on the full knowledge of the joint distribution of (X, Y, Z). Results are reported in Table 5.2. In particular, bias-corrected Cramer's V (cf. [4]); Theil's U and AIC are calculated by considering each of the available X.

TABLE 5.1
Description of the variables used in the simulation.

Name	Description
cntry	Country of residence (20 countries)
hhmmb5	No. of Household members (1, 2, 3, 4, >4)
gndr	Gender of respondent (1="Male", 2="Female")
ageacat	Classes of age of respondent (14-17, 18-24, 25-44, 45-64, 65-74, >74)
icpart1	Respondent lives with husband/wife/partner (1="Yes", 2="No")
chldhm	Children in the household (1="Yes", 2="No")
anypacc	Problems with accommodation (1="Yes", 2="No")
eisced	Highest level of education (7 categories)
prof2	Professional status (1="In paid work", 2="Retired", 3="Other")
crpdwk	Done paid work (≥1 hour) in last 7 days (those with prof2=1 or 2)
emplrel	Type of work (those with prof2=1) (1="Employee", 2="Self-empl.", 3="In family's business")
hinctnta	Deciles of household total income (1-10)
hincfel1	Living comfortably on present income (1="Yes", 2="No")

TABLE 5.2

Association, PRE measures and AIC between $W = (Y \times Z)$ and each of the available X.

X	V'	U	AIC
cntry	0.1117	0.0409	158818.1
hhmmb5	0.2380	0.0393	158507.0
gndr	0.0966	0.0018	164575.4
ageacat	0.1312	0.0153	162495.1
icpart1	0.3878	0.0267	160463.2
chldhm	0.2539	0.0119	162901.5
anypacc	0.1479	0.0038	164231.6
eisced	0.1785	0.0328	159653.6
prof2	0.2805	0.0278	160321.4
crpdwk	0.2606	0.0251	160771.1
emplrel	0.0688	0.0025	164522.7

The bias-corrected V indicated "icpart1" as the variable with highest association with the combination of the target variables ($W = Y \times Z$), followed by "prof2", "crpdwk", "chldhm" and "hhmmb5" showing lower association values. Slightly different subset of variables can be identified by considering Theil's U or AIC, in particular: "cntry", "hhmmb5", "eisced" and, with smaller values, "icpart1", "prof2", and "crpdwk". When AIC is used to identify the best combination of X predicting $W = Y \times Z$, a subset with just the variables "hhmmb5" and "cntry" is returned as first choice (combination with minimum AIC), followed by ("hhmmb5", "eisced") and ("hhmmb5", "cntry", "icpart1"). These results show that even in the ideal case in a unique data source having all the variables (X, Y, Z), pairwise measures, not taking into account relationships between predictors, suggest to select a higher number of predictors than those necessary.

Now consider the whole sample but keeping separate the two target variables, to show how basic analyses work when the two distinct samples are available, as in a standard SM application. This represents the ideal situation in SM where the bivariate distributions (X, Y) and (X, Z) are known.

All the measures (see Table 5.3) point to "cntry" and "eisced" as best predicts of "hincfel1" (Y). On the contrary, results are not fully coherent when the response variable is "hinctnta" (Z); inspection of U and AIC estimates determine the choice of a larger subset of matching variables ("hhmmb5", "icpart1", "eisced", "prof2", "crpdwk") if compared to the one that V' would indicate ("icpart1", "prof2", "crpdwk").

Assuming that $X_Y = $ ("cntry", "eisced") and $X_Z = $ ("hhmmb5", "icpart1", "eisced", "prof2", "crpdwk"), their intersection determine a subset consisting of just a single matching variable, $X_M = $ "eisced"; on the contrary, their union returns a set of six matching variables ("cntry", "hhmmb5", "icpart1",

TABLE 5.3
Association, PRE measures and AIC for all possible combinations between each X, Y and Z, respectively.

	Response="hincfel1" (Y)			Response="hinctnta" (Z)		
X	V'	U	AIC	V'	U	AIC
cntry	0.3567	0.1027	33427.9	0.0898	0.0168	130381.1
hhmmb5	0.0960	0.0073	36948.6	0.2260	0.0436	126565.4
gndr	0.0488	0.0019	37144.6	0.0915	0.0019	132020.9
ageacat	0.0411	0.0014	37170.3	0.1149	0.0150	130367.3
icpart1	0.1008	0.0080	36916.6	0.3849	0.0327	127949.7
chldhm	0.0529	0.0022	37132.2	0.2077	0.0097	130994.3
anypacc	0.1096	0.0100	36844.0	0.1285	0.0035	131815.9
eisced	0.2385	0.0438	35593.3	0.1715	0.0376	127395.2
prof2	0.1343	0.0146	36672.0	0.2716	0.0326	127981.4
crpdwk	0.1143	0.0103	36834.4	0.2596	0.0308	128211.3
emplrel	0.0564	0.0026	37120.1	0.0663	0.0028	131943.9

"eisced", "prof2", "crpdwk"). It is easy to see that the joint distribution of these variables would produce a table with 12,600 cells, which doubles when Y is added, and become 126,000 for $Z \times X_M$, too many to be reliably estimated with a sample of 28,769 ($\tilde{n}$, the average cell frequency, would be 0.2288, well below the threshold of 1). However, the top 3 subsets of combinations of predictors of joint $Y \times Z$ identified by AIC are all subsets of the larger one achieved by union of best predictors of Y and Z, separately considered.

Using AIC-based procedure to identify the "optimal" combination of predictors for each response variable, separately considered, it comes out that $X_Y^{(AIC)}$ = ("cntry", "eisced", "prof2") and $X_Z^{(AIC)}$ = ("hhmmb5", "icpart1", "eisced", "prof2"). The intersection consists of two matching variables X_M = ("eisced", "prof2") while the union gives a set with five predictors ("cntry", "hhmmb5", "icpart1", "eisced", "prof2"), slightly smaller from the one achieved after application of pairwise measures. Again, the best 3 combinations of predictors of $Y \times Z$ identified by AIC are subsets of the latter X_M.

Selecting the matching variables by exploring uncertainty is computationally demanding but has the advantage of avoiding separate analyses. It requires the estimation of three contingency tables: joint distribution of the X variables, joint distributions of $X \times Y$ and $X \times Z$. Having a unique sample corresponds to an ideal situation because the tables estimated from it will not show discrepancies in the estimated joint distributions of the X variables, as it happens when dealing with two independent samples.

We have selected the matching variables by using the algorithm introduced in Section 5.2.4 according to the two alternative penalised measures $\tilde{d}_m^{(1)}$ and $\tilde{d}_m^{(2)}$. The results are shown in Tables 5.4 and 5.5. The procedure for selecting automatically the matching variables based on penalized

TABLE 5.4
Automatic selection of matching variables with penalised measure $\bar{d}_m^{(1)}$.
Tables estimated from the whole sample.

Comb. of X (D_m)	No. of X	H_{D_m}	$\bar{d}_m^{(1)}$	$\bar{n}$
unconditional	0	2	0.100000	2876.90
cntry	1	20	0.098797	143.85
cntry eisced	2	140	0.093235	20.55
cntry eisced hhmmb5	3	700	0.087449	4.11
cntry eisced hhmmb5 prof2	4	2100	0.079922	1.37

TABLE 5.5
Automatic selection of matching variables with $\bar{d}_m^{(2)}$. Tables estimated from
the whole sample.

Comb. of X (D_m)	No. of X	H_{D_m}	$\bar{d}_m^{(2)}$	$\bar{n}$
unconditional	0	2	0.100000	2876.90
cntry	1	20	0.098829	143.85
cntry eisced	2	140	0.093248	20.55
cntry eisced hhmmb5	3	700	0.087380	4.11
cntry eisced hhmmb5 prof2	4	2100	0.079704	1.37

uncertainty provides the same results whatever the penalty is. The second
penalty term seems to be slightly higher that the first one, but this does
not affect the intermediate and final results. In fact, in both examples the
identified set of best matching variables consists of just 4 of the 11 starting
common variables: "cntry", "hhmmb5", "eisced", and "prof2". This subset
is smaller (4 vs. 6 variables) than the one that is achieved by basic separate
analyses when considering the union of best predictors of the target variables ($X_Y \cup X_Z$) but slightly larger than ("cntry", "hhmmb5") identified as
"optimal" by AIC when predicting $Y \times Z$.

It is worthwhile to remark that these results refer to an "ideal" case in
which in practice all the variables are available, i.e., a case in which there
is no uncertainty and statistical matching is not needed. To figure out what
would happen in a matching setting, the whole sample is considered as an
artificial population that serves as a basis for simulations. At each iteration:
(1) Two independent simple random samples, A and B, are drawn without
replacement from it. Variable Z ("hinctnta") is removed from A while Y
("hincfel1") is removed from B.
(2) Automatic selection of matching variables is performed using traditional
methods (pairwise measures V' and AIC) and the two uncertainty-based procedures but with a different penalty term.
(3) The various identified subsets of matching variables are saved.

TABLE 5.6

Subset of matching variables identified in the simulation according to the traditional selection methods.

Measure	Chosen X (matching variables)	(a)	(b)	(c)	(d)
V'	cntry crpdwk eisced hhmmb5 icpart1 prof2	100	100	100	100
U	cntry crpdwk eisced hhmmb5 icpart1 prof2	61	70	13	2
	cntry crpdwk eisced hhmmb5 prof2	7	6	8	10
	cntry eisced hhmmb5 icpart1 prof2	30	24	79	88
	Others	2			
AIC p.	cntry eisced hhmmb5 icpart1	55	53	76	61
	cntry eisced hhmmb5 prof2	31	35	22	39
	cntry hhmmb5 icpart1 prof2	8	8	2	
	Others	6	4		
AIC c.	cntry hhmmb5 icpart1 prof2	10	6		
	cntry eisced icpart1 prof2	21	85	100	99
	cntry eisced icpart1	34			
	cntry hhmmb5 prof2	24	3		
	cntry icpart1 prof2	5			
	Others	6	6		1

A set of 100 iterations of this procedure are carried out for each of the following combinations of n_A and n_B: (a) $n_A = 1000$ and $n_B = 3000$; (b) $n_A = 2000$ and $n_B = 4000$; (c) $n_A = 5000$ and $n_B = 10000$; (d) $n_A = 10000$ and $n_B = 15000$.

Table 5.6 shows results of simulations of traditional procedures for selecting matching variables; in particular, results refer to the union of variables with highest association or being best predictor of Y and Z, respectively. Pairwise measures gives variable results just in case (a), i.e., with smallest sizes of samples, with the exception of V', that identifies always a subset of 6 matching variables, the same identified in the analysis on the full sample. When Theil's U is considered, a subset composed of 5 variables is preferred to the larger ones only in cases (c) and (d). The penalty in AIC, denoted by "AIC p". in Table 5.6, tends to favor smaller subsets of matching variables; the most frequent ones differ just on one of the four variables. Finally, when the matching variables are determined by union of the best combination of predictors of Y and Z, respectively, as identified to AIC, denoted by "AIC c". in Table 5.6, results show a very high heterogeneity in case (a), showing that this procedure suffers of too small samples if compared to the number of cells to be estimated. From case (b) onwards, heterogeneity disappears and the four variables set ("cntry", "eisced", "hhmmb5", "prof2") is the one preferred in almost all the simulations.

In summary, the small set of simulations shows that traditional methods for selecting matching variables tend to provide wider sets of variables than the one strictly needed. Just AIC-based selection procedures, taking into

TABLE 5.7
Number of times a set of matching variables is selected in the different simulations cases.

Case	Selected matching variables	Freq.	$\sqrt{(MSE(\hat{d})/d^*}$
(a)	chldhm cntry eisced	19	0.02%
	cntry eisced gndr	23	0.02%
	cntry eisced icpart1	57	0.02%
	cntry gndr hhmmb5	1	
(b)	chldhm cntry eisced	8	0.01%
	cntry eisced gndr	6	0.01%
	cntry eisced icpart1	86	0.01%
(c)	ageacat cntry eisced	1	
	cntry eisced hhmmb5	99	0.01%
(d)	ageacat cntry eisced	1	
	cntry eisced gndr hhmmb5	99	0.00%

account a penalty related to the number of parameters to be estimates, tend to identify smaller sets of matching variables. As expected, high heterogeneity of results characterizes case with relatively small samples ($n_A = 1000$ and $n_B = 3000$). In general, the sets of matching variables being identified with traditional selection procedures tend to be larger that the 'best' combination identified when working in the ideal situation (all the variables are available).

Let's now examine results of simulations (in Table 5.7) related to the procedure of selecting matching variables based on uncertainty. For each combination of sample sizes, there is the tendency to reproduce with higher frequencies in the 100 iterations a specific combination of matching variables (the higher the sample sizes, the higher the frequencies, in italics in the table). Just for the combination $n_A = 5000$ and $n_B = 10000$, the selected matching variables are coherent with the analyses performed on the whole dataset (in bold and italic in the table); the other combinations include also other variables (as "icpart1" or "gndr"). Probably this is due to the fact that multivariate associations are the most difficult aspects to estimate. Anyway, once a choice is made, the MSE of $\hat{d}$ is under control and its relative square root with respect to the reference known d^* tends to zero sharply. In the simulation setting, d^* is d calculated on the whole ESS sample considered as the target population from which the various samples are selected.

As expected, the uncertainty-based procedures for identifying matching variables are more parsimonious if compared to traditional ones based on pairwise measures, even compared to those based on AIC which also include a penalty term.

5.4 Conclusions

This work gives an overview of the approaches to the problem of selecting matching variables in a statistical matching application. This is a very time-consuming phase when analyzing large-scale surveys. Two ways of working are compared as being very different in terms of both simplicity and computational effort. In our case, dealing with categorical variables, a very frequent case with household surveys, the traditional procedures based on pairwise association/PRE measures are very simple and require a very low computational effort but tend to provide too many matching variables, mainly when Cramer's V or Theil's U are considered. Smaller sets are instead selected when looking at AIC, because of the penalty term that favors parsimonious models. The smaller and effective subsets of matching variables are those identified by procedures based on investigation of uncertainty including a penalty term, these procedures are, however, computational demanding. In general, the set of matching variables identified by means of uncertainty tends to be a subset of the one identified with traditional pairwise measures; this outcome suggests that in the presence of data sources sharing a high number of common variables X a two-step procedure could be used: at the first step, traditional measures are used to identify a subset of most relevant common variables then, in the second step, uncertainty exploration can be used to identify the set of matching variables starting from the subset of X identified in the first step. This complex procedure would join benefits of both procedures, i.e., simplicity of the traditional procedure and parsimony of the one based on uncertainty, with a reduced computational effort due to applying investigation of the uncertainty starting from a smaller subset of the possible variables X.

Bibliography

[1] Agresti, A. (2002), Categorical Data Analysis, 2nd Edition. Wiley, New York.

[2] Agresti, A., M.C. Yang (1987), An empirical investigation of some effects of sparseness in contingency tables, Computational Statistics & Data Analysis, 5, 9–21.

[3] Anderson, T.W. (1957), Maximum likelihood estimates for a multivariate normal distribution when some observations are missing, Journal of the American Statistical Association, 52, 200–203.

[4] Bergsma, W. (2013), A bias-correction for Cramer's V and Tschuprow's T. Journal of the Korean Statistical Society, 42, 323-328.

[5] Bishop, Y.M., S.E. Fienberg, P.W. Holland (1975), Discrete Multivariate Analysis: Theory and Practice. M.I.T. Press, Cambridge, MA.

[6] Breiman, L., J.H. Friedman, R.A. Olshen, C.J. Stone (1984), Classification and Regression Trees. Wadsworth.

[7] Breiman, L. (2001), Random Forests, Machine Learning 45(1), 5-32.

[8] Cohen, M.L. (1991), Statistical matching and microsimulation models. In: Citro, Hanushek (eds) Improving Information for Social Policy Decisions: The Uses of Microsimulation Modeling. Vol II Technical papers, Washington D.C.

[9] Conti, P.L., D. Marella, M. Scanu (2012), Uncertainty analysis in statistical matching, Journal of Official Statistics, 28(1), 69–88.

[10] Conti, P.L., D. Marella, M. Scanu (2016), Statistical matching analysis for complex survey data with applications, Journal of the American Statistical Association, 111(516), 1715–1725.

[11] Conti, P.L., D. Marella, M. Scanu (2017), How far from identifiability? A systematic overview of the statistical matching problem in a non-parametric framework. Communications in Statistics – Theory and Methods, 46(2), 967–994.

[12] Conti, P.L., D. Marella, M. Scanu (2019), An overview on uncertainty and estimation in statistical matching. In Zhang L.-C., Chambers, R. Analysis of integrated data, CRC/Chapman and Hall.

[13] D'Orazio, M. (2016), StatMatch: Statistical matching (aka data fusion). R package version 1.2.4.
https://CRAN.R-project.org/package=StatMatch

[14] D'Orazio, M. (2016), Statistical matching and imputation of survey data with StatMatch. R package Vignette.
https://CRAN.R-project.org/package=StatMatch

[15] D'Orazio, M., M. Di Zio, M. Scanu (2006), Statistical matching for categorical data: Displaying uncertainty and using logical constraints, Journal of Official Statistics, 22(1), 1–22.

[16] D'Orazio, M., M. Di Zio, M. Scanu (2006), Statistical Matching: Theory and Practice. Wiley, Chichester.

[17] Di Zio, M., B. Vantaggi (2017), Partial identification in statistical matching with misclassification, International Journal of Approximate Reasoning, 82(3), 227–241.

[18] Harrell Jr, F.E. (2015), Regression Modeling Strategies With Applications to Linear Models, Logistic and Ordinal Regression, and Survival Analysis, 2nd Edition. New York: Springer.

[19] Imbens, G., C.F. Manski (2004), Confidence intervals for partially identified parameters. Econometrica, 72, 1845–1857.

[20] Manski, C. (1995), Identification Problems in the Social Sciences. Harvard University Press.

[21] Moriarity, C., F. Scheuren (2001), Statistical matching: A paradigm for assessing the uncertainty in the procedure, Journal of Official Statistics, 17, 407–422.

[22] Quine, S., quoted in Aitkin, M. (1978), The analysis of unbalanced cross classifications (with discussion). Journal of the Royal Statistical Society series A, 141, 195–223.

[23] Rässler, S. (2002), Statistical Matching: A frequentist theory, practical applications and alternative Bayesian approaches. Lecture Notes in Statistics, New York: Springer Verlag.

[24] Rubin, D.B. (1974), Characterizing the estimation of parameters in incomplete-data problems, Journal of the American Statistical Association, 69(346), 467–474.

[25] Rubin, D.B. (1986), Statistical matching using file concatenation with adjusted weights and multiple imputations, Journal of Business and Economic Statistics, 4(1), 87–94.

[26] Sakamoto, Y., H. Akaike (1978), Analysis of Cross-Classified bata by AIC. Annals of the Institute of Statistical Mathematics, 30, B, 185–197.

[27] Särndal, C.E., B. Swensson, J. Wretman (1992), Model Assisted Survey Sampling. Springer, New York.

[28] Stoye, J. (2009), More on confidence intervals for partially identified parameters. Econometrica, 77, 1299–1315.

[29] Strobl, C., A.L. Boulesteix, A. Zeileis, T. Hothorn (2007), Bias in Random Forest Variable Importance Measures: Illustrations, Sources and a Solution. BMC Bioinformatics, 8: 25.

[30] Tamer, E. (2010), Partial identification in econometrics, Annual Review of Economics 2(1), 167–195.

[31] Vantaggi, B. (2008), Statistical matching of multiple sources: A look through coherence, International Journal of Approximate Reasoning, 49(3), 701–711.

[32] Zhang, L.C. (2015), On proxy variables and categorical data fusion, Journal of Official Statistics, 31(4), 783–807.

[33] Zhang, L.C., R.L. Chambers (2019), Minimal inference from incomplete 2x2-tables. In Zhang L.-C., Chambers, R. Analysis of integrated data, CRC/Chapman and Hall.

[34] ESS (2014), European Social Survey Round 7 Data (2014). Data file edition 2.1. NSD - Norwegian Centre for Research Data, Norway – Data Archive and distributor of ESS data for ESS ERIC.

6

Minimal inference from incomplete 2 × 2-tables

Li-Chun Zhang

University of Southampton, Statistics Norway & University of Oslo

Raymond L. Chambers

University of Wollongong, Wollongong, Australia

CONTENTS

6.1 Introduction

Incomplete 2×2 tables are often encountered in statistical analysis. Table 6.1 illustrates the two cases that we pay special attention to in this chapter. Both tables correspond to the cross-classification of two binary variables. To the left, $X = 1, 0$ is the outcome variable of interest, and $R = 1, 0$ indicates whether an observation is missing or not. The two-way table is incomplete since X is only observed if $R = 1$. We refer to it as the *missing data setting*. The two-way table on the right shows the joint distribution of two binary variables X and Y. This table is completely unobserved. Instead, one has observations on two independent samples of n_1 values of X and n_2 values of Y, respectively. We refer to it as the *matched data setting*. For either case, we assume that the *complete data* corresponding to the unobserved 2×2 table follow a multinomial distribution P_λ, with parameter $\lambda = (\lambda_{11}, \lambda_{10}, \lambda_{01}, \lambda_{00})$ referring to the probabilities of observing each of the four possible configurations of two binary variables. Under this assumption, the parameter λ is *point-identifiable* given the complete data, in the sense that $\lambda = \lambda'$ whenever

121

TABLE 6.1

Two cases of incomplete 2×2-table. Left: binary variable subjected to missing data, sample size n; right: statistical matching of two binary variables from separate samples of sizes n_1 and n_2, respectively. Unobserved hypothetical complete sample counts marked by '−'.

Hypothetical Complete Sample Data: $(n_{11}, n_{01}, n_{10}, n_{00}) \sim$ multinomial$(n, \lambda_{11}, \lambda_{01}, \lambda_{10}, \lambda_{00})$

	R=1	R=0	Total
X=1	n_{11}	−	−
X=0	n_{01}	−	−
Total	n_{+1}	n_{+0}	n

Observation: (n_{11}, n_{01}, n_{+0}) Given n

Sampling Distribution: multinomial$(n, \lambda_{11}, \lambda_{01}, \lambda_{+0})$

Identifiable: $\lambda_{11}, \lambda_{01}, \lambda_{+0} = \lambda_{10} + \lambda_{00}$

Parameter of Interest: $\theta = \lambda_{1+}$

Identification Region: $\lambda_{11} \leq \theta \leq \lambda_{11} + \lambda_{+0}$

Additional Assumption: Independent (X, R)

	Y=1	Y=0	Total
X=1	−	−	n_x
X=0	−	−	$n_1 - n_x$
Total	n_y	$n_2 - n_y$	

Observation: Independent $(n_x, n_1), (n_y, n_2)$

Sampling Distribution: $n_x \sim$ binomial(n_1, λ_{1+}); $n_y \sim$ binomial(n_2, λ_{+1})

Identifiable: $\lambda_{1+} = \lambda_{11} + \lambda_{10}; \lambda_{+1} = \lambda_{11} + \lambda_{01}$

Parameter of Interest: $\theta = \lambda_{11}$

Identification Region: $\theta \leq \min(\lambda_{1+}, \lambda_{+1})$; $\theta \geq \max(\lambda_{1+} + \lambda_{+1} - 1, 0)$

Additional Assumption: Independent (X, Y)

$P_\lambda = P_{\lambda'}$; and so is the parameter of interest $\theta = \Pr(X = 1)$ in the missing data setting and $\theta = \Pr(X = 1, Y = 1)$ in the matched data setting.

Table 6.1 also shows the sampling distribution of the observed data for each setting. Point-identification for θ based on the observed data is only achievable if additional assumptions are made. In this table, these assumptions are independence of X and R in the missing data setting, which means missing-completely-at-random (MCAR, Rubin, 1976), and independence of X and Y in the statistical matching setting, which is a special case of the conditional independence assumption (Okner, 1972). But such additional assumptions are often contentious. It therefore seems reasonable to ask 'what the data say' about θ given the *accepted* sampling distribution of the observed data, *without* the additional "esoteric" (Tamer, 2010) assumptions that enable point-identification of this parameter. Our aim here is to describe a general approach to inference based on incomplete 2×2 tables given such a setting.

To illustrate, consider a missing data example discussed by Zhang (2010). The observed data from the Obstructed Coronary Bypass Graft Trials (OCBGT, see Hollis, 2002) are $(n_{11}, n_{01}, n_{+0}) = (32, 54, 24)$, with the sampling distribution parameter $\psi = (\lambda_{11}, \lambda_{01}, \lambda_{+0})$. The likelihood of ψ is proportional to $\lambda_{11}^{n_{11}} \lambda_{01}^{n_{01}} \lambda_{+0}^{n_{+0}}$. This yields the profile likelihood of the parameter of interest $\theta = \lambda_{1+}$, denoted by $L_p(\theta)$, which is the dashed curve in Figure 6.1. It is seen that $L_p(\theta)$ is flat over $[n_{11}/n, (n_{11} + n_{+0})/n]$, which we call the *maximum likelihood region*, denoted by $\widehat{\Theta}$, with all values of θ in $\widehat{\Theta}$ equally likely based on the observed data. Asymptotically, as $n \to \infty$, $\widehat{\Theta}$ tends to the *identification region* of θ, i.e., $\lambda_{11} \le \theta \le \lambda_{11} + \lambda_{+0}$, which is a function of the identifiable parameter ψ. This identification region is the asymptote of 'what the data say' about θ under the setting here. The dotted curve gives the standardised likelihood under the additional MCAR assumption that enables point-identification of θ. It peaks at the maximum likelihood estimate (MLE) $\widehat{\theta}_{MCAR} = n_{11}/n_{+1}$, which converges to $\lambda_{11}/\lambda_{+1}$ in probability. Clearly, the MLE derived from the MCAR likelihood will be *inconsistent* as long as $\lambda_{1+} \ne \lambda_{11}/\lambda_{+1}$.

The fact that the profile likelihood shown in Figure 6.1 is constant within the observed $\widehat{\Theta}$ does not mean that all the values of θ in it are equally likely to be in a $\widehat{\Theta}$ that could be observed given a random draw from the sampling distribution of the observed data. In Section 6.2 we develop the concept of *corroboration*, noting that values of θ that are more likely to appear in a $\widehat{\Theta}$ on repeated sampling are better corroborated by the observed data than values of θ that only infrequently appear in a $\widehat{\Theta}$. The solid curve in Figure 6.1 shows how the estimated corroboration varies with θ for the OCBGT data. The computation of the estimated corroboration is explained in Section 6.2. The key point to note here is that the corroboration varies for the points within $\widehat{\Theta}$, where the profile likelihood is constant. This allows us to construct high corroboration level sets *within* $\widehat{\Theta}$. It will be shown that asymptotically the set of values with the maximum observed corroboration becomes

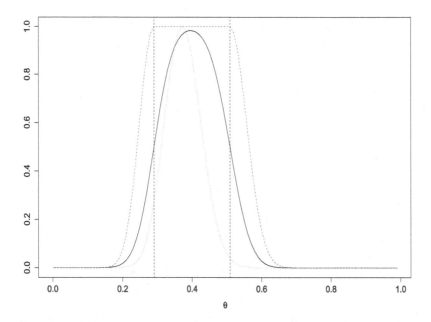

FIGURE 6.1
Observed corroboration and standardised likelihoods (with peak value 1) based on OCBGT data: observed corroboration (solid), profile likelihood (dashed), likelihood under MCAR assumption (dotted); maximum likelihood region marked by vertical dashed lines.

indistinguishable from the identification region except for its bounds. Unlike the MLE that aims at the *most likely* parameter value, the maximum corroboration set identifies those parameter values that are the *hardest to refute* based on the observed data. In effect, these are the points in which we have the highest confidence. We develop a Corroboration Test in Section 6.5 for the settings of Table 6.1, where the Likelihood Ratio Test is inapplicable insofar as the parameter of interest is not point identifiable. The test will be applied to the OCBGT data.

There are several related approaches within the matched data setting. In ecological inference (Goodman, 1953; King, 1997), the observed data are the margins of the unobserved complete 2×2 table. See Wakefield (2004) for a comprehensive review. It is clearly recognised that critical but untestable assumptions are needed to arrive at a point estimate in this context, and that there is a fundamental difficulty associated with choosing between different models based only on the observed data; see Greenland and Robins (1994), Freedman (2001) and Gelman *et al.* (2001). Statistical matching deals with the same setting, where the set of multinomial distributions P_λ compatible with the sampling distribution of the observed data is referred to as the *uncertainty space*. Evaluation of the uncertainty space has received much attention (Kadane, 1978; Moriarity and Scheuren, 2001; D'Orazio *et al.*, 2006; Kiesel and Rässler, 2006; Conti *et al.* 2012; Zhang, 2015; Conti *et al.*, 2015). The concept of uncertainty space is closely related to that of identification uncertainty (Koopmans, 1949; Tamer, 2010). The "partial identification" framework (Manski, 1995, 2003, 2007) recognises situations where, due to the structure of the data, even a hypothetical infinite number of observations may only constrain the parameter of interest without being able to point-identify it. It is important in this context to distinguish between the study of identification, provided an *infinite* amount of data under the given structure, and statistical inference from *finite* samples. Partial identification in econometrics can be traced back to Frisch (1934) and Marschak and Andrews (1944), and there is a growing literature on the construction of confidence regions of the identified parameter set. See e.g., Imbens and Manski (2004), Chernozhukov *et al.* (2007), Beresteanu and Molinari (2008), and Romano and Shaikh (2010).

It is clear that all the aforementioned approaches aim at inference, based on an identifiable sampling distribution that is acceptable to all, no matter which untestable additional assumptions an analyst may or may not introduce in order to resolve the identification issue. As seen in Figure 6.1, the novelty of the approach proposed in the sequels is that it achieves this objective via a measure of the statistical evidence in the observed data that is *not* based on comparing likelihoods.

6.2   Corroboration

Denote by $f(d_n; \psi)$ the identifiable sampling distribution of the observed data d_n with generic sample size n, and with parameter ψ. Denote by P_λ the distribution of the hypothetical complete data, which is characterised by the parameter λ with parameter space Λ. Denote by $\theta = \theta(\lambda)$ a scalar parameter of interest, and by Θ the parameter space of θ. For any given ψ let $\Lambda(\psi)$ be the constrained parameter space defined by ψ. That is, $\Lambda(\psi)$ consists of all λ that are consistent with ψ. Let $\Theta(\psi)$ be the induced parameter space of θ, which contains all $\theta(\lambda)$ where $\lambda \in \Lambda(\psi)$. For inference under a *minimal setting* in this chapter, we then require both conditions below to hold.

(**M₁**) The induced parameter space $\Theta(\psi)$ is a closed interval. In particular, it is not a singleton $\Theta(\psi) = \theta(\psi)$, nor is it invariant towards ψ in the sense that $\Theta(\psi) = \Theta(\psi')$ for all $\psi \neq \psi'$.

(M₂) The parameter ψ of the sampling distribution is point-identifiable, and the MLE $\widehat{\psi}$ is such that $\widehat{\psi} \overset{\mathrm{Pr}}{\to} \psi_0$, asymptotically as $n \to \infty$, where ψ_0 is the true parameter value.

Under a minimal setting, $\Theta(\psi) = [L(\psi), U(\psi)]$, where $L(\psi)$ is the *lower bound* of θ induced by ψ, and $U(\psi)$ the *upper bound*. The identification region is $\Theta_0 = \Theta(\psi_0) = [L_0, U_0]$, where $L_0 = L(\psi_0)$ and $U_0 = U(\psi_0)$. Thus, for the missing data setting in Table 6.1, we have $\psi_0 = (\lambda_{11}^0, \lambda_{01}^0, \lambda_{+0}^0)$, with

$$\Theta_0 = [L_0, U_0] = [\lambda_{11}^0, \ \lambda_{11}^0 + \lambda_{+0}^0].$$

For the matched data setting, we have $\psi_0 = (\lambda_{1+}^0, \lambda_{+1}^0)$, and the Fréchet bounds (Fréchet, 1951) define the identification region

$$\Theta_0 = [L_0, U_0] = [\max(\lambda_{1+}^0 + \lambda_{+1}^0 - 1, 0), \ \min(\lambda_{1+}^0, \lambda_{+1}^0)].$$

Let $\widehat{L} = L(\widehat{\psi})$ and $\widehat{U} = U(\widehat{\psi})$ be the MLEs of L_0 and U_0, respectively, and let $\widehat{\Theta} = \Theta(\widehat{\psi}) = [\widehat{L}, \widehat{U}]$ denote the maximum profile likelihood estimator of θ. The points inside $\widehat{\Theta}$ can all be considered as *equally most likely*, i.e., best supported according to the likelihood based on d_n under the observed data model. We define the *corroboration function* of θ, for $\theta \in \Theta$, to be

$$c(\theta; \psi) = \Pr(\theta \in \widehat{\Theta}; \psi), \tag{6.1}$$

i.e., the probability for the given value of θ to be covered by $\widehat{\Theta}$, where the probability is evaluated with respect to $f(d_n; \psi)$. Let the *actual corroboration* be

$$c_0(\theta) = c(\theta; \psi_0),$$

i.e., evaluated over the true sampling distribution. In particular, $c(\theta_0; \psi_0)$ is the confidence level of $\widehat{\Theta}$ as an interval estimator of θ_0. Let the *observed corroboration* be

$$\widehat{c}(\theta) = c(\theta; \widehat{\psi}).$$

Since $\widehat{c}(\theta)$ is the MLE of $c_0(\theta)$, one may then define the observed corroboration as the *most likely level of corroboration* for θ given the observed data. As illustrated in Figure 6.1 for the OCBGT data, if one treats the observed corroboration as a function of θ, then this function can generally vary over $\widehat{\Theta}$, as opposed to the profile likelihood which is flat over the same region. Note that in this case in order to calculate $\widehat{c}(\theta)$, where $(\widehat{\lambda}_{11}, \widehat{\lambda}_{+0}) = (n_{11}/n, n_{11}/n + n_{+0}/n)$, we employ the bivariate normal approximation $(\widehat{\lambda}_{11}, \widehat{\lambda}_{+0}) \sim N_2(\mu, \Sigma)$, where $\mu = (\lambda_{11}, \lambda_{+0})$ and the distinctive elements of Σ are $V(\widehat{\lambda}_{11}) = \lambda_{11}(1 - \lambda_{11})/n$, $V(\widehat{\lambda}_{+0}) = \lambda_{+0}(1 - \lambda_{+0})/n$ and $Cov(\widehat{\lambda}_{11}, \widehat{\lambda}_{+0}) = -\lambda_{11}\lambda_{+0}/n$. More generally, the observed corroboration can be calculated via simulation as follows.

Bootstrap for $\widehat{c}(\theta)$

For given θ and the MLE $\widehat{\psi}$, repeat for $b = 1,...B$:

- generate $d_n^{(b)}$ from $f(d_n; \widehat{\psi})$ to obtain $\widehat{\psi}^{(b)}$ and the corresponding $[L(\widehat{\psi}^{(b)}), U(\widehat{\psi}^{(b)})]$;

- set $\delta^{(b)} = 1$ if $\theta \in [L(\widehat{\psi}^{(b)}), U(\widehat{\psi}^{(b)})]$, and 0 otherwise.

Put $\widehat{c}(\theta) = \sum_{b=1}^{B} \delta^{(b)}/B$ as the bootstrap estimate of the observed corroboration for θ. $\square$

6.3 Maximum corroboration set

Let the *level-α corroboration set* be given by

$$A_\alpha(\psi) = \{\theta : c(\theta; \psi) \geq \alpha\},$$

provided there exists some $\theta \in A_\alpha(\psi)$ where $c(\theta; \psi) = \alpha$. Thus, by definition we have $c(\theta; \psi) < \alpha$, for any $\theta \notin A_\alpha(\psi)$, whilst we *cannot* have $c(\theta; \psi) > \alpha$ for all $\theta \in A_\alpha(\psi)$. Some properties of $A_\alpha(\psi)$ are given below. Notice that we use $c(\theta)$ as a short-hand for $c(\theta; \psi)$ and A_α that of $A_\alpha(\psi)$, where it is not necessary to emphasise their dependence on ψ.

Theorem 1 *Suppose that a minimal inference setting applies, i.e., provided conditions* (M_1) *and* (M_2) *hold. (i) Let* $A_{\alpha_1} = [L_1, U_1]$ *and* $A_{\alpha_2} = [L_2, U_2]$. *If* $\alpha_1 > \alpha_2$, *then* $[L_1, U_1] \subset [L_2, U_2]$. *(ii) Let* $\theta_L < \theta_U$, *where* $c(\theta_L) = c(\theta_U) = \alpha$. *Then* $c(\theta) \geq \alpha$ *for any* $\theta \in (\theta_L, \theta_U)$.

Proof *(i) On the one hand, we have* $A_{\alpha_1} \setminus A_{\alpha_2} = \emptyset$ *because, otherwise, there must exist some* $\theta \in A_{\alpha_1} \setminus A_{\alpha_2}$ *such that* $c(\theta) \geq \alpha_1$ *(because* $\theta \in A_{\alpha_1}$*) and* $c(\theta) < \alpha_2$ *(because* $\theta \notin A_{\alpha_2}$*) at once, contradictory to* $\alpha_1 > \alpha_2$ *as stipulated. On the other hand, the set* $A_{\alpha_2} \setminus A_{\alpha_1}$ *is non-empty because, otherwise, every* $\theta \in A_{\alpha_2}$ *must belong to* A_{α_1} *and, thus,* $c(\theta) \geq \alpha_1$, *so that there exists no* $\theta \in A_{\alpha_2}$ *such that* $c(\theta) = \alpha_2 < \alpha_1$, *contradictory to the definition of* A_{α_2}.

(ii) Each $\widehat{\Theta}$ *can be classified into 4 distinct types, denoted by (a)* $\widehat{\Theta}_{L\bar{U}}$ *where* $\theta_L \notin \widehat{\Theta}$ *and* $\theta_U \notin \widehat{\Theta}$, *(b)* $\widehat{\Theta}_{LU}$ *where* $\theta_L \in \widehat{\Theta}$ *and* $\theta_U \in \widehat{\Theta}$ *and, thus,* $\theta \in \widehat{\Theta}_{LU}$, *(c)* $\widehat{\Theta}_L$ *where* $\theta_L \in \widehat{\Theta}$ *and* $\theta_U \notin \widehat{\Theta}$, *(d)* $\widehat{\Theta}_U$ *where* $\theta_L \notin \widehat{\Theta}$ *and* $\theta_U \in \widehat{\Theta}$. *Type (c) can be further classified into (c.1)* $\widehat{\Theta}_{L1}$ *where* $\theta \in \widehat{\Theta}_{L1}$ *and (c.2)* $\widehat{\Theta}_{L2}$ *where* $\theta \notin \widehat{\Theta}_{L2}$, *i.e., depending on whether or not* θ *appears in* $\widehat{\Theta}$. *Similarly, type (d) into (d.1)* $\widehat{\Theta}_{U1}$

where $\theta \in \widehat{\Theta}_{U1}$ and (d.2) $\widehat{\Theta}_{U2}$ where $\theta \notin \widehat{\Theta}_{U2}$. We have

$$c(\theta_L) = \Pr(\widehat{\Theta}_{LU}) + \Pr(\widehat{\Theta}_L) = \Pr(\widehat{\Theta}_{LU}) + \Pr(\widehat{\Theta}_{L1}) + \Pr(\widehat{\Theta}_{L2})$$
$$c(\theta_U) = \Pr(\widehat{\Theta}_{LU}) + \Pr(\widehat{\Theta}_U) = \Pr(\widehat{\Theta}_{LU}) + \Pr(\widehat{\Theta}_{U1}) + \Pr(r\widehat{\Theta}_{U2})$$
$$c(\theta) \geq \Pr(\widehat{\Theta}_{LU}) + \Pr(\widehat{\Theta}_{L1}) + \Pr(\widehat{\Theta}_{U1}).$$

Thus, if $\Pr(\widehat{\Theta}_{U1}) \geq \Pr(\widehat{\Theta}_{L2})$, then $c(\theta) \geq c(\theta_L)$, or if $\Pr(\widehat{\Theta}_{U1}) \leq \Pr(\widehat{\Theta}_{L2})$, then $\Pr(\widehat{\Theta}_{L1}) \geq \Pr(\widehat{\Theta}_{U2})$ since $c(\theta_L) = c(\theta_U)$, such that $c(\theta) \geq c(\theta_U)$. Similarly on comparison between $\Pr(\widehat{\Theta}_{L1})$ and $\Pr(\widehat{\Theta}_{U2})$. □

Theorem 2 *Given a minimal inference setting, there exists a maximum corroboration value denoted by $\theta^{\max}$, such that $c(\theta^{\max}) \geq c(\theta)$ for any $\theta \neq \theta^{\max}$.*

Proof *Take any initial level-α_1 corroboration set $A_{\alpha_1} = [L_{\alpha_1}, U_{\alpha_1}]$. Without losing generality, one of the end points must have corroboration α_1 by Theorem 1.i; suppose $c(L_{\alpha_1}) \geq c(U_{\alpha_1}) = \alpha_1$. By definition $c(\theta) \geq \alpha_1$ for all $\theta \in A_{\alpha_1}$. If $c(\theta) = c(L_{\alpha_1})$ for all $L_{\alpha_1} < \theta < U_{\alpha_1}$, then $\theta^{\max} = L_{\alpha_1}$, since $c(\theta) < \alpha_1 \leq c(L_{\alpha_1})$ for any $\theta \notin A_{\alpha_1}$. Otherwise, there exists $L_{\alpha_1} < \theta < U_{\alpha_1}$, where $c(\theta) = \alpha_2 > c(L_{\alpha_1}) \geq \alpha_1$, and the corresponding level-α_2 corroboration set, denoted by $A_{\alpha_2} = [L_{\alpha_2}, U_{\alpha_2}]$. By Theorem 1.i, we have $[L_{\alpha_2}, U_{\alpha_2}] \subset [L_{\alpha_1}, U_{\alpha_1}]$. Since $\alpha \leq 1$, iteration of the argument must terminate at some maximum level-α. □*

Denote by $A^{\max} = A^{\max}(\psi_0)$ the *maximum corroboration set*, such that $c_0(\theta) > c_0(\theta')$ for any $\theta \in A^{\max}$ and $\theta' \notin A^{\max}$, and $c_0(\theta) = c_0(\theta')$ for any $\theta \neq \theta' \in A^{\max}$. It follows from (6.1) that these are the points for which $\widehat{\Theta}$ implies the highest confidence, in which sense one may consider these to be the parameter values that are the hardest to refute. Replacing ψ_0 by $\widehat{\psi}$, we obtain the MLE of $A^{\max}$ or the *observed* maximum corroboration set

$$\widehat{A}^{\max} = A^{\max}(\widehat{\psi}).$$

Figure 6.2 illustrates corroboration in the matched data setting, where $\theta = \lambda_{11}$. The true sampling distribution parameters $(\lambda_{1+}, \lambda_{+1})$ are $(0.1, 0.9)$ for the left plot and $(0.3, 0.3)$ to the right. The sample sizes are $(n_1, n_2) = (1000, 500)$ to the left and $(200, 300)$ to the right. The identification region Θ_0 is the interval between the vertical dashed lines, and the solid curve shows how the actual corroboration (denoted *cvalue* in the plots) varies with θ. The corroboration of some interior points of Θ_0 can be 1, whereas it can be 0 for many $\theta \notin \Theta_0$. In the left plot, both $c_0(L_0)$ and $c_0(U_0)$ are about 0.5; in the right plot, we have $c_0(L_0) = 1$ and $c_0(U_0) \approx 0.25$.

Let $\bar{c}(\theta; \psi) = \lim_n c(\theta; \psi) = \lim_n \Pr(\theta \in \widehat{\Theta}_n; \psi)$ be the *asymptotic corroboration* of θ evaluated at ψ, where $\lim_n$ stands for $\lim_{n \to \infty}$ and $\widehat{\Theta}_n$ makes explicit the dependence on sample size. Table 6.2 summarises the asymptotic actual corroboration $\bar{c}_0(\theta) = \bar{c}(\theta; \psi_0)$ for both data settings. Let $\bar{A}^{\max}$ be the asymptotic maximum actual corroboration set based on $\bar{c}_0(\theta)$. Lemma 1 states that,

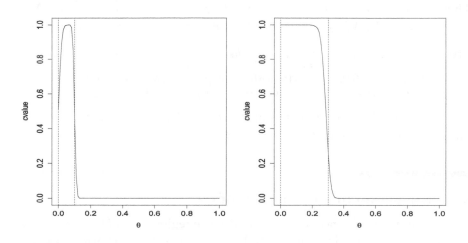

FIGURE 6.2
Illustration of corroboration in matched data setting. Left: $(\lambda_{1+}, n_1) = (0.1, 1000)$ and $(\lambda_{+1}, n_2) = (0.9, 500)$. Right: $(\lambda_{1+}, n_1) = (0.3, 200)$ and $(\lambda_{+1}, n_2) = (0.3, 300)$.

TABLE 6.2
Asymptotic actual corroboration $\bar{c}_0(\theta)$ in missing and matched data settings

Data Setting	$\theta \notin [L_0, U_0]$	$\theta = L_0$	$\theta \in (L_0, U_0)$	$\theta = U_0$
Missing	0	0.5 if $L_0 > 0$	1	0.5 if $U_0 < 1$
Matching	0	0.5 if $\lambda_{1+} + \lambda_{+1} \geq 1$ 1 if $\lambda_{1+} + \lambda_{+1} < 1$	1	0.5 if $\lambda_{1+} \neq \lambda_{+1}$ 0.25 if $\lambda_{1+} = \lambda_{+1}$

apart from the bounds L_0 and U_0, $\bar{A}^{\max}$ is indistinguishable from Θ_0 and $\bar{c}_0(\theta)$ is an indicator function on Θ_0. Theorem 3 states that the interior of the observed maximum corroboration set $\widehat{A}_n^{\max}$ converges to the interior of Θ_0 in probability.

Lemma 1 *Given a minimal inference setting, $\theta \in \bar{A}^{\max}$ and $\bar{c}_0(\theta) = 1$ if $\theta \in Int(\Theta_0) = (L_0, U_0)$, i.e., if θ belongs to the interior of Θ_0, then $\theta \notin \bar{A}^{\max}$ and $\bar{c}_0(\theta) = 0$ for any $\theta \notin [L_0, U_0]$.*

Proof *Let $\delta(\theta; \widehat{\psi}_n) = 1$ if $\theta \in Int(\widehat{\Theta}) = (\widehat{L}_n, \widehat{U}_n)$, and 0 otherwise, where $\widehat{\psi}_n$ is the MLE. Without losing generality, for any $\theta = U_0 - \epsilon$, where $0 < 2\epsilon < U_0 - L_0$, we have $\delta(\theta; \widehat{\psi}_n) = 1$ if $|\widehat{U}_n - U_0| < \epsilon$ and $|\widehat{L}_n - L_0| < \epsilon$, the probability of which tends to 1, since $\widehat{\psi}_n \xrightarrow{Pr} \psi_0$. Thus, $\delta(\theta; \widehat{\psi}_n) \xrightarrow{Pr} 1$, i.e., $\bar{c}_0(\theta) = 1$ and $\theta \in \bar{A}^{\max}$. Similarly, it can be shown that $\bar{c}_0(\theta) = 1$, for $\theta \notin \Theta_0$, i.e., $\theta \notin \bar{A}^{\max}$.* $\square$

Theorem 3 *Given a minimal inference setting, we have* $Int(\widehat{A}^{\max}) \overset{Pr}{\to} Int(\Theta_0)$; *that is,* $\lim_n Pr(\theta \in \widehat{A}_n^{\max}) = 1$ *if* $\theta \in Int(\Theta_0)$ *and* $\lim_n Pr(\theta \in \widehat{A}_n^{\max}) = 0$ *if* $\theta \notin \Theta_0$.

Proof *By the general form of Slutsky's Theorem (e.g., Theorem 7.1, Kapadia et al., 2005), we have* $\bar{c}(\theta; \widehat{\psi}_n) \overset{Pr}{\to} \bar{c}(\theta; \psi_0)$, *since* $\widehat{\psi}_n \overset{Pr}{\to} \psi_0$ *and* $\bar{c}(\theta; \psi)$ *is a bounded for all* ψ. *Thus, if* $\theta \in (L_0, U_0)$, *such that* $\bar{c}(\theta; \psi_0) = 1$ *by Lemma 1, we have* $\bar{c}(\theta; \widehat{\psi}_n) \overset{Pr}{\to} \bar{c}(\theta; \psi_0) = 1$, *meaning* $\lim_n Pr(\theta \in \widehat{A}_n^{\max}) = 1$. *Similarly, it can be shown that* $\lim_n Pr(\theta \in \widehat{A}_n^{\max}) = 0$, *for* $\theta \notin \Theta_0$. $\square$

6.4 High assurance estimation of Θ_0

Given a minimal inference setting, a confidence region C_n for Θ_0 (which is an interval) has the confidence level $Pr(\Theta_0 \subseteq C_n)$; see, e.g., Chernozhukov *et al.* (2007). Given a high confidence level, the probability that C_n contains points that do not belong to Θ_0 must also be high, due to sampling variability, and so C_n asymptotically contracts towards Θ_0 from 'outside' of it. In contrast, any point in Θ_0 is irrefutable, and $\widehat{A}^{\max}$ identifies those parameter values that are the hardest to refute given the observed data. We thus define the *assurance* of $\widehat{A}^{\max}$ to be

$$\tau_0 = Pr(\widehat{A}^{\max} \subseteq \Theta_0),$$

where the probability is evaluated with respect to $f(d_n; \psi_0)$. That is, this is the probability that the points in the observed $\widehat{A}^{\max}$ are indeed all irrefutable. If $\widehat{A}^{\max}$ has a high assurance, there will be a low probability that it contains points outside of Θ_0. As the sample size increases, a high assurance estimator of Θ_0 should therefore grow towards Θ_0 from 'inside' of it. In light of Theorem 1, for some small constant $h \geq 0$, a high assurance estimator of Θ_0 can therefore be defined as

$$\widehat{A}_h = \{\theta : c(\theta; \widehat{\psi}) \geq \max_\theta c(\theta; \widehat{\psi}) - h\}.$$

The following bootstrap can be used to estimate $\widehat{A}_h$, including $\widehat{A}_0 = \widehat{A}^{\max}$.

Bootstrap for $\widehat{A}_h$

Given the MLE $\widehat{\psi}$ and the corresponding $[\widehat{L}, \widehat{U}]$, repeat for $b = 1, ... B$:

1. generate $d_n^{(b)}$ from $f(d_n; \widehat{\psi})$, and obtain $\widehat{\psi}^{(b)}$;

2. for any given h, where $0 \leq h < 1$, obtain $\widehat{A}_h^{(b)}$ at $\widehat{\psi}^{(b)}$ in the same way as $\widehat{A}_h$ at $\widehat{\psi}$, and the corresponding $L^{(b)} = L(\widehat{A}_h^{(b)})$ and $U^{(b)} = U(\widehat{A}_h^{(b)})$;

3. set $\delta^{(b)} = 1$ if $\widehat{L} \leq L^{(b)} < U^{(b)} \leq \widehat{U}$, and $\delta^{(b)} = 0$ otherwise.

Calculate the bootstrap estimate of assurance as $\widehat{\tau}(\widehat{A}_h; \psi_0) = \sum_{b=1}^{B} \delta^{(b)}/B$, with corresponding bootstrap estimate of the lower end of Θ_0 given by $L(\widehat{A}_h) = \sum_{b=1}^{B} L^{(b)}/B$ and of the upper end of Θ_0 given by $U(\widehat{A}_h) = \sum_{b=1}^{B} U^{(b)}/B$. $\square$

For small h, $\widehat{A}_h$ can have higher assurance than $\widehat{\Theta}$, whereas it can be 'closer' to $\bar{A}^{\max}$ than $\widehat{A}^{\max} = \widehat{A}_0$ by Theorem 1, since $\widehat{A}_0 \subset \widehat{A}_h$. Setting $h < 0.25$ makes $\text{Int}(\widehat{A}_h)$ asymptotically indistinguishable from $\text{Int}(\Theta_0)$ for the two settings depicted in Table 6.2. In a finite-sample situation, one may calculate $\widehat{A}_h$ and its assurance for several different choices of h. Since the length of $\widehat{A}_h$ increases with h while its assurance decreases, one may choose the longest $\widehat{A}_h$ as an estimator of Θ_0 subject to an acceptable level of assurance.

6.5 A corroboration test

Consider testing the null hypothesis $H_A : \theta^* \in (L_0, U_0)$ against $H_B : \theta^* \notin \Theta_0$. A minimal inference setting for this test is non-standard because, under both H_A and H_B, the set of possible distributions of the observed data are exactly the same, i.e., $f(d_n; \psi)$. The Likelihood Ratio Test is inapplicable. Let instead the test statistic be $T_n = 1$ if $\theta^* \in \text{Int}(\widehat{\Theta}_n)$ and $T_n = 0$ if $\theta^* \notin \widehat{\Theta}_n$. Suppose we reject H_A if $T_n = 0$. The power function of this testing procedure is then $\beta_n(\theta^*) = \Pr(T_n = 0; \psi_0)$, and is such that

$$\bar{\beta}(\theta^*) \equiv \lim_n \beta_n(\theta^*) = 1 - \lim_n \Pr(T_n = 1; \psi_0) = 1 - \bar{c}_0(\theta^*).$$

If H_A is true, but $T_0 = 0$ and we reject H_A, by Lemma 1 the probability of Type-I error converges to zero since $\bar{c}_0(\theta^*) = 1$ if $\theta \in \text{Int}(\Theta_0)$. Similarly, if H_B is true, but $T = 1$ and we do not reject H_A, the Type-II error probability also asymptotes to zero since $\bar{c}_0(\theta^*) = 0$ if $\theta^* \notin \Theta_0$.

TABLE 6.3
Supporting evidence for $H_A : \theta^* \in (L_0, U_0)$ vs. $H_B : \theta^* \notin \Theta_0$.

	Low Power $\widehat{\beta}_n(\theta^*)$	High Power $\widehat{\beta}_n(\theta^*)$
$T_n = 1$	Support H_A	Support neither, improbable event
$T_n = 0$	Support neither, improbable event	Support H_B

Let the *observed power* be $\widehat{\beta}_n(\theta^*) = 1 - \widehat{c}_n(\theta^*)$, which is a consistent estimator of $\bar{\beta}(\theta^*)$. While $\widehat{c}_n(\theta^*)$ is a consistent estimator of the Type-II error probability, we cannot use it to estimate the Type-I error probability. The reason is that $c_0(\theta^*)$ is the same under H_A or H_B, due to the minimal inference setting, so that it cannot be related to *both* types of errors. We shall therefore define the *Corroboration Test* to have observed power β, where $\beta = \widehat{\beta}_n(\theta^*) \in (0,1)$, if H_A is rejected when $T_n = 0$. As summarised in Table 6.3, a Corroboration Test of high observed power would lead one to reject θ^* if it is outside of $\widehat{\Theta}_n$ and have a low observed corroboration. By the consistency of $\widehat{A}_n^{\max}$ established in Theorem 3, we have

$$\lim_n \Pr(\text{Reject } H_A \text{ when } H_A \text{ is true}) = 0 < \lim_n \Pr(\text{Reject } H_A \text{ when } H_B \text{ is true}) = 1.$$

That is, the Corroboration Test is strongly Chernoff-consistent (Theorem 4), since T_n has limiting size 0 and the Type-II error probability converges to 0, for any θ^* specified in H_A.

Theorem 4 *Given a minimal inference setting, the Corroboration Test of observed power* $\beta = \widehat{\beta}_n(\theta^*)$, *for* $\beta \in (0,1)$, *is strongly Chernoff-consistent.*

6.6   Application: missing OCBGT data

Consider the OCBGT data $n = (n_{11}, n_{01}, n_{+0}) = (32, 54, 24)$. The profile likelihood is

$$L_p(\theta) \propto \begin{cases} n_{11}^{\theta} n_{01}^{\frac{(1-\theta)n_{01}}{n_{01}+n_{+0}}} n_{+0}^{\frac{(1-\theta)n_{+0}}{n_{01}+n_{+0}}} & \text{if } \theta < \widehat{\lambda}_{11} \\[2ex] n_{11}^{\widehat{\lambda}_{11}} n_{01}^{\widehat{\lambda}_{01}} n_{+0}^{\widehat{\lambda}_{+0}} & \text{if } \widehat{\lambda}_{11} \leq \theta \leq \widehat{\lambda}_{11} + \widehat{\lambda}_{+0} \\[2ex] n_{11}^{\frac{\theta n_{11}}{n_{11}+n_{+0}}} n_{01}^{1-\theta} n_{+0}^{\frac{\theta n_{+0}}{n_{11}+n_{+0}}} & \text{if } \theta > \widehat{\lambda}_{11} + \widehat{\lambda}_{+0} \end{cases}$$

(Zhang, 2010). The likelihood is $L_{MCAR}(\theta) \propto n_{11}^{\theta} n_{01}^{1-\theta}$, under the additional assumption of independent (X, R). Figure 6.1 plots both, as well as the observed corroboration $\widehat{c}(\theta)$.

The likelihood L_{MCAR} does not vary with n_{+0}, e.g., whether this value is 4, 24 or 104. Accordingly n_{+0} is not part of the available statistical evidence. Clearly, such insensitiveness towards the observed data requires some external belief to sustain. Next, consider the relative plausibility of $\theta^* = 0.2, 0.3, 0.5, 0.6$ against $\theta_1 = 0.4$ based on the profile likelihood ratio, denoted by $LR_p(\theta^*, \theta_1)$ in the left part of Table 6.4. The values 0.3 and 0.5 cannot be distinguished from 0.4, since all are inside $\widehat{\Theta} = [0.29, 0.51]$; the negative evidence of 0.2 and 0.6 against 0.4 is "moderate" according to Royall (1997), as they fall in the range $1/32 - 1/8$. Nevertheless, as noted before, the Likelihood Ratio Test is inapplicable here.

Now, based on the observed corroboration $\widehat{c}(\theta^*)$ in Table 6.4, one may reject the null hypothesis $H_0 : 0.2 \in \Theta_0$ on the basis of the Corroboration Test with observed power 0.982. Similarly for $H_0 : 0.6 \in \Theta_0$, with observed

TABLE 6.4
Left, profile likelihood ratio $LR_p(\theta^*, \theta_1)$ with $\theta_1 = 0.4$, observed corroboration $\widehat{c}(\theta^*)$ based on OCBGT data. Right, assurance $\widehat{\tau}(\widehat{A_h}; \psi_0)$ of $\widehat{A_h}$, expected left end $L(\widehat{A_h})$ and right end $U(\widehat{A_h})$, with values obtained by bootstrap with $B = 5000$. In addition, $\widehat{\Theta} : [\widehat{L}, \widehat{U}] = [0.29, 0.51]$, $\widehat{\tau}(\widehat{\Theta}) = 0.19$.

θ^*	$LR_p(\theta^*, \theta_1)$	$\widehat{c}(\theta^*)$	h	$\widehat{\tau}(\widehat{A_h}; \psi_0)$	$[L(\widehat{A_h}), U(\widehat{A_h})]$
0.2	0.076	0.018	0	0.99	$[0.40, 0.40]$
0.3	1	0.583	0.01	0.95	$[0.38, 0.41]$
0.4	1	0.985	0.06	0.84	$[0.36, 0.44]$
0.5	1	0.576	0.40	0.25	$[0.30, 0.50]$
0.6	0.156	0.028	0.80	0.00	$[0.25, 0.55]$

power 0.972. Meanwhile, 0.3 and 0.5 are just inside $\widehat{\Theta}$, with $\widehat{c}(0.3)$ and $\widehat{c}(0.5)$ slightly below 0.6, and so cannot be rejected with high observed power. The Corroboration Test thus allows us to reject an unlikely value of θ with a high observed power.

Finally, five observed corroboration level sets $\widehat{A_h}$ are illustrated in the right part of Table 6.4, where the estimated assurance $\widehat{\tau}(\widehat{A_h}; \psi_0)$ and expected end points $L(\widehat{A_h})$ and $U(\widehat{A_h})$ are calculated using the bootstrap described in Section 6.4. As an estimator of Θ_0, $\widehat{A_0}$ is very narrow but has 99% assurance; $\widehat{A_{0.01}}$ has 95% assurance and is expected to span from 0.38 to 0.41. Using $\widehat{\Theta}$ as an estimator of Θ_0 would perform comparably to $\widehat{A_{0.4}}$, but with low assurance. The observed corroboration level sets thus allow us to identify true irrefutable points in Θ_0 with a high assurance.

Bibliography

[1] Beresteanu, A. and Molinari, F. (2008). Asymptotic properties of a class of partially identified models. *Econometrica*, **76**, 763-814.

[2] Chambers, R.L. and Steel, D. (2001). Simple methods for ecological inference in 2×2 tables. *J. R. Statist. Soc. A*, **164**, 175-192.

[3] Chao, A. (1987). Estimating the population size for capture-recapture data with unequal catchability. *Biometrics*, **43**, 783-791.

[4] Chernozhukov, V., Hong, H., Tamer, E. (2007). Estimation and confidence regions for parameter sets in econometric models. *Econometrica*, **75**, 1243-1284.

[5] Conti, P.L., Marella, D. and Scanu, M. (2015). Statistical matching analysis for complex survey data with applications. *J. Am. Statist. Ass.*, DOI: 10.1080/01621459.2015.1112803

[6] Conti, P.L., Marella, D. and Scanu, M. (2012). Uncertainty analysis in statistical matching. *J. Off. Statist.*, **28**, 69-88.

[7] D'Orazio, M., Di Zio, M. and Scanu, M. (2006). *Statistical Matching: Theory and Practice*. Chichester: Wiley.

[8] Freedman, D. A. (2001). Ecological inference and the ecological fallacy. In *International Encyclopaedia of the Social and Behavioural Sciences* (eds N. J. Smelser and P. B. Baltes), vol. 6, pp. 4027-4030. New York: Elsevier.

[9] Fréchet, M. (1951). Sur les tableaux de correlation dont les marges sont données. *Ann. Univ. Lyon A*, **3**, 53-77.

[10] Frisch, R. (1934). *Statistical Confluence Analysis*. Publ. No. 5. Oslo: Univ. Inst. Econ.

[11] Gelman, A., Park, D. K., Ansolabehere, S., Price, P.N. and Minnite, L. C. (2001). Models, assumptions and model checking in ecological regressions. *J. R. Statist. Soc. A*, **164**, 101-118.

[12] Goodman, L. (1953). Ecological regressions and the behavior of individuals. *Am. Sociol. Rev.*, **18**, 663-666.

[13] Greenland, S. and Robins, J. (1994). Ecological studies: biases, misconceptions and counterexamples. *Am. J. Epidem.*, **139**, 747-760.

[14] Hollis, S. (2002). A graphical sensitivity analysis for clinical trials with nonignorable missing binary outcome. *Stat. Med.*, **21**, 3823-3834.

[15] Imbens, G. and Manski, C.F. (2004). Confidence intervals for partially identified parameters. *Econometrica*, **72**, 1845-1857.

[16] Kadane, J.B. (1978). Some Statistical Problems in Merging Data Files. In *1978 Compendium of Tax Research*, pp. 159-171. U.S. Department of Treasury. (Reprinted in J. Off. Statist., **17**, 423-433.)

[17] Kapadia, A.S., Chan, W. and Moyé, L. (2005). *Mathematical Statistics with Applications*. Chapman & Hall/CRC.

[18] Kiesl, H. and Rässler, S. (2006). *How valid can data fusion be?* Institut fur Arbeitsmarkt- und Berufsforschung (IAB) Discussion Paper 15/2006.

[19] King, G. (1997). *A Solution to the Ecological Inference Problem: Reconstructing Individual Behavior from Aggregate Data*. Princeton: Princeton University Press.

[20] Koopmans, T. (1949). Identification problems in economic model construction. *Econometrica*, **17**, 125-144.

[21] Manski, C.F. (1995). *Identification Problems in the Social Sciences.* Harvard University Press.

[22] Manski, C.F. (2003). *Partial Identification of Probability Distributions.* New York: Springer.

[23] Manski, C.F. (2007). *Identification for Prediction and Decision.* Cambridge, MA: Harvard Univ. Press.

[24] Marschak, J. and Andrews, W.H. (1944). Random simultaneous equations and the theory of production. *Econometrica*, **12**, 143-203.

[25] Moriarity, C. and Scheuren, F. (2001). Statistical matching: A paradigm for assessing the uncertainty in the procedure. *J. Off. Statist.*, **17**, 407-422.

[26] Nadarajah, S. and Kotz, S. (2008). Exact distribution of the max/min of two Gaussian random variables. *IEEE Transactions on very large scale integration (VLSI) systems*, **16**(2), 210-212.

[27] Okner, B.A. (1972). Constructing a new microdata base from existing microdata sets: the 1966 merge file. *Ann. Econ. Soc. Mea.*, **1**, 325-342.

[28] Romano, J. and Shaikh, A. (2010). Inference for the identified set in partially identified econometric models. *Econometrica*, **78**, 169-211.

[29] Royall, R. (1997). *Statistical Evidence: A Likelihood Paradigm.* Chapman & Hall.

[30] Rubin, D.B. (1976). Inference and missing data. *Biometrika*, **63**, 581-592.

[31] Tamer, E. (2010). Partial identification in econometrics. *Annu. Rev. Econ.*, **2**, 167-195.

[32] Wakefield, J. (2004). Ecological inference for 2×2 tables. (With discussions). *J. Roy. Statist. Soc. A*, **167**, 385-445.

[33] Zhang, Z. (2010). Profile likelihood and incomplete data. *Int. Statist. Rev.*, **78**, 102-116.

[34] Zhang, L.-C. (2015). On proxy variables and categorical data fusion. *J. Off. Statist.*, **31**, 783-807.

7

Dual- and multiple-system estimation with fully and partially observed covariates

Peter G. M. van der Heijden

University of Southampton, Southampton, UK, and Utrecht University, Utrecht, The Netherlands

Paul A. Smith

University of Southampton, Southampton, UK

Joe Whittaker

University of Lancaster, Lancaster, UK

Maarten Cruyff

Utrecht University, Utrecht, The Netherlands

Bart F. M. Bakker

Statistics Netherlands, Den Haag, The Netherlands, and VU University, The Netherlands

CONTENTS

7.1   Introduction

There are many situations in which we want to know the size of a population; however, the information about that population is imperfect. A well-known technique for estimating the size of a human population is to find two or more data sources or registers which identify members of the population of interest, to link the individuals in the registers and use the linked information to estimate the number of individuals that occur in neither of the registers (Fienberg [10], Bishop, Fienberg and Holland [4], Cormack [7], International Working Group for Disease Monitoring and Forecasting, IWGDMF [15]; see also the Chapters 8 and 9 in this volume by di Cecco [8] and Zhang [31]). For example, with two registers A and B, linkage gives a count of individuals in A but not in B, a count of individuals in B but not in A, and a count of individuals both in A and B. The counts form a contingency table denoted by $A \times B$ with the variable labeled A being short for 'inclusion in register A', taking the levels 'yes' and 'no', and likewise for register B. In this table the cell 'no, no' has a zero count by definition, and the statistical problem is to estimate the size of the population of interest in this cell. This estimate is added to the counts of individuals found in at least one of the registers, giving an improved population-size estimate relative to using only numbers of people observed in at least one of the registers.

For dual- and multiple-system estimation with two or more sources, the usual assumptions under which a population-size estimate is obtained are

- the population is closed

- individuals in registers A and B can be linked perfectly

- there is no overcoverage due to individuals that are not part of the target population

and, in addition, with two registers:

- inclusion in register A is independent of inclusion in register B; and

- in at least one of the two registers, the inclusion probabilities are homogeneous (see Chao et al. [6], Zwane, van der Pal-de Bruin and van der Heijden [35] and van der Heijden et al. [27]). Interestingly it is often, but incorrectly, supposed that the inclusion probabilities in *both* registers have to be homogeneous.

However, it is generally agreed that these assumptions are unlikely to hold in human populations. Three approaches may be adopted to make the impact of possible violations less severe.

One approach is to include categorical covariates in the model, in particular covariates whose levels have heterogeneous inclusion probabilities for both registers (see Bishop, Fienberg and Holland [4] and Baker [2] for approaches in human populations, and compare with Pollock [20] who reviews applications in animal populations). Then log-linear models can be fitted to the higher-way contingency table of registers A and B and the covariates. The restrictive independence assumption in two-source estimation is replaced by a less restrictive assumption of independence of A and B conditional on the covariates; and subpopulation-size estimates are derived (one for every level of the covariates) that add up to a population-size estimate. Including covariates in log-linear models of population registers improves population-size estimates for two reasons. Firstly, it takes account of heterogeneity of inclusion probabilities over the levels of the covariate; and secondly, it subdivides the estimated population (and hence the estimate of the unobserved part of the population) by the levels of the covariates, giving insight into characteristics of individuals that are not included in any of the registers.

A second approach is to include a third register, and to analyse the resulting three-way contingency table with log-linear models that may include one or more two-factor interactions, thus modelling the dependence and avoiding the need for an independence assumption. The (less stringent) assumption that the three-factor interaction is absent is needed, because this interaction cannot be estimated from the observed data. Including a third register is not always possible, as it is not available, or because there is no information that makes it possible to link the individuals in the third register to both the other registers.

A third approach makes use of a latent variable to take heterogeneity of inclusion probabilities into account (see Fienberg, Johnson and Junker [11], Bartolucci and Forcina [3]). These authors use the Rasch model, a model often used in psychometrics. The model assumes a continuous latent variable, and given the latent variable, the inclusion probabilities are independent. A lower position on the latent variable causes all inclusion probabilities to be lower and a high position causes all inclusion probabilities to be higher. For example, if the data concern four different medical registers, the latent variable could have an interpretation as severity of the illness, where patients with a higher severity have inclusion probabilities that are higher.

Of course, these three approaches are not exclusive and may be used concurrently in one model. When the approach adopted includes using covariates, there is a question of which covariates should be chosen. In the traditional approach, only covariates that are available in all of the registers can be chosen. We refer to these as fully observed covariates, and describe approaches in Sections 7.2–7.3. In Sections 7.4–7.5 the focus is on covariates that are available in only some of the registers (see Zwane and van der Heijden [34] and van der Heijden et al. [27]), and we refer to these as incomplete covariates. Whether or not the covariates are available in each of the registers, the number of possible log-linear models that can be fitted grows rapidly as the number of sources and/or covariates increases. In Section 7.6 we introduce an application in which conceptually the same variable is measured in two registers, but because the measurement processes are different it is better to treat them as different variables, and this has possible applications in analysing consistency of similar variables in different sources. Finally in Section 7.7 we give a general discussion of the issues raised.

7.2 Theory concerning invariant population-size estimates

In this section we study the (in)variance of population-size estimates derived from log-linear models that include covariates, with the same covariates available in all the registers. First, we need to set up some notation and terminology.

7.2.1 Terminology and properties

We adopt a dual notation where inclusion in the two registers A and B is also coded by variables A and B, with levels $A, B = 0, 1$ where level 0 refers to not registered, and we assume that there are I categorical covariates denoted by X_i, where $i = 1, \ldots, I$. The contingency table classified by variables A, B and X_1 is denoted by $A \times B \times X_1$. We denote hierarchical log-linear models by their highest fitted margins using the notation of Bishop, Fienberg and Holland [4]. For example, in the absence of covariates, the independence model is denoted by $[A][B]$, and when there is one covariate X_1 the model with A and B conditionally independent given X_1 is $[AX_1][BX_1]$. In each of these models, the two-factor interaction between A and B is absent, which reflects the (conditional) independence assumption discussed in the Introduction, and that the two-factor interaction cannot be fitted without additional information.

The number of independent parameters in the model cannot be greater than the number of observed counts. When the number of parameters is equal to the number of observed counts, we have a *saturated* model, and in this case the fitted values from the model are equal to the observed counts. The table $A \times B$ has a single structural zero, so there are three observed counts and the saturated model is $[A][B]$ (with parameters for the intercept, A and B). When there are I covariates, the saturated model for the table $A \times B \times X_1 \times \cdots \times X_I$ is $[AX_1 \ldots X_I][BX_1 \ldots X_I]$, where A and B are conditionally independent given the covariates.

From any multi-way contingency table, it is possible to produce a smaller table by summing over one of the margins (and therefore eliminating this margin from the table), and this is called *marginalising* the table. For example, starting with contingency table $A \times B \times X_1$, if we marginalise over X_1 we obtain the table $A \times B$.

When the full table of registers by covariates can be marginalised over one or more covariates without changing the population-size estimate we will say that the table (or the model) is *collapsible* over these covariates. By focussing on population-size estimates, collapsibility in log-linear models is studied in this paper from a different perspective than found in Bishop, Fienberg and Holland [4], who are interested in parametric collapsibility. Our work applies the model collapsibility of Asmussen and Edwards [1], later discussed by Whittaker [29] and Kim and Kim [16], concerning the commutativity of model fitting and marginalisation. Applying the idea of collapsibility to a table containing structural zeros is a novel extension first suggested in van der Heijden et al. [27].

We then distinguish between passive covariates, where marginalising a model over the covariate does not affect the population-size estimate, and active covariates, which cannot be removed from the model without changing the population-size estimate. While passive covariates do not affect the size estimate, which suggests that they might be ignored, a secondary objective of population-size estimation is to provide estimates of the size of sub-populations, or equivalently, to break down the population size in terms of given covariates, which may well include passive covariates. Creating a population breakdown by passive covariates through an appropriate collapsible model is an elegant way to tackle this important practical problem.

The property of collapsibility in tables and the associated marginalisation of models have two closely related corollaries that we wish to examine:

1. There exist log-linear models for which the table is collapsible over specific covariates.

2. For a given contingency table there exist different log-linear models that yield identical total population-size estimates.

The corollaries are closely related because if corollary 2 applies to a contingency table, then it is collapsible over the covariates which are not common to the two models. We first illustrate the corollaries in a specific example and then provide an explanation.

7.2.2   Example

We take an example from the quality evaluation of the population census in the Netherlands in 2011, where the population of interest is people aged 15-64 with Afghan, Iranian or Iraqi nationality; for full details, see van der Heijden et al. [27]. In this case A is the Dutch population register (GBA), and B is the police register (HKS), and we use two covariates, X_1 denoting the gender of the person and X_2 denoting their age coded in four bands (15-25, 25-35, 35-50 and 50-64). Table 7.1 gives the estimates from a range of models which can be fitted to the resulting four-way table.

Table 7.1 illustrates corollary 1, that some models allow for collapsing a table over specific covariates. For example, model M_1 produces the same population-size estimate as M_0, and therefore under M_1 the $A \times B \times X_1$ table is collapsible over X_1. Table 7.1 also shows corollary 2, that the same estimates may be obtained from a range of different models – six different models yield an estimate of 6,170.3. Indeed we can go further with corollary 2, since there

TABLE 7.1
Log-linear model specifications, fit statistics and the resulting population-size estimates (for the population not observed in either register) from the range of models which can be fitted to the $A \times B \times X_1 \times X_2$ table taking account of the structural zeroes.

model id	model definition	deviance	df	AIC	population size estimate
models fitted to the $A \times B$ (sub)table					
M_0	$[A][B]$	0.0	0	28.9	6,170.3
models fitted to the $A \times B \times X_1$ (sub)table					
M_1	$[AX_1][B]$	548.5	1	558.5	6,170.3
M_2	$[A][BX_1]$	1.1	1	11.1	6,170.3
M_3	$[AX_1][BX_1]$	0.0	0	12.0	5,696.1
models fitted to the $A \times B \times X_1 \times X_2$ table					
M_4	$[AX_1][BX_2]$	617.6	13	639.6	6,170.3
M_5	$[AX_1][BX_1][X_2]$	228.6	15	246.6	5,696.1
M_6	$[AX_1X_2][B]$	718.2	7	752.2	6,170.3
M_7	$[AX_1][AX_2][X_1X_2][B]$	725.6	10	753.6	6,170.3
M_8	$[AX_1][BX_2][X_1X_2]$	588.6	10	616.6	6,179.4
M_9	$[AX_1][BX_1][BX_2]$	69.1	12	93.1	5,696.1
M_{10}	$[AX_1][BX_1][X_1X_2]$	200.2	12	224.2	5,696.1
M_{11}	$[AX_1][BX_2][AX_2][BX_1]$	65.9	9	95.9	5,837.1
M_{12}	$[AX_1][BX_1X_2]$	4.9	6	40.9	5,696.1
M_{13}	$[AX_1][BX_1][BX_2][[X_1X_2]$	34.4	9	64.4	5,696.1
M_{14}	$[AX_1X_2][BX_1X_2]$	0.0	0	48.0	5,910.1
M_{15}	$[AX_1X_2][BX_1][BX_2]$	23.3	3	65.3	6,257.1
M_{16}	$[AX_1][AX_2][BX_1][BX_2][X_1X_2]$	31.2	6	67.2	5,831.4

are invariant (identical) population-size estimates which result from two different kinds of data reduction:

1. marginalisation or collapsing, which yields a lower-dimension table,

2. specification of sufficient margins and a corresponding unsaturated hierarchical model based on the original table.

For example, in Table 7.1 the largest table is $A \times B \times X_1 \times X_2$, and all the models M_0, M_1, M_2, M_4, M_6 and M_7 yield the same population estimate. M_0 is based on the table $A \times B$ by marginalisation over X_1 and X_2; M_1 and M_2 are unsaturated models based on the table $A \times B \times X_1$ by marginalisation over X_2 only; and M_4, M_6 and M_7 are unsaturated models based on the table $A \times B \times X_1 \times X_2$, that is without marginalisation at all. These latter models are however collapsible, since they produce the same population estimates as simpler models, and therefore X_2 is a passive covariate in these models. It is interesting to note that it is not only nested models which produce the same population estimates – for example, model M_4 is not nested within M_6 or M_7, because it has a BX_1 interaction.

From this we see that corollary 1 implies corollary 2 (whenever the table is collapsible across covariates, there is a corresponding reduced model), and that corollary 2 implies corollary 1, since the same estimate may only be produced by models that are collapsible.

Table 7.2 shows the observed population counts and the fitted population estimates from four of the models. Note that, even though the total population-size estimates for models M_1 and M_2 are equal, estimates of the subpopulation sizes (for males and females) from M_1 are different from those under M_2. Under model M_2 the fitted values are much closer to the observed values than under model M_1, and this illustrates the two deviances for these models, which are 1.1 for model M_2 and 548.5 for model M_1, see Table 7.1.

TABLE 7.2

Observed cases and estimates from models M_0, M_1, M_2 and M_3 for the three-way table of A (GBA), B (HKS) and X_1 (gender); for A and B level 1 is present and for X_1 level 1 is male

A	B	X_1	obs	M_0	M_1	M_2	M_3
1	1	1	972	1,085	629.2	976.5	972.0
1	1	0	113		455.8	108.5	113.0
0	1	1	234	26,254	234.0	229.5	234.0
0	1	0	21		21.0	25.5	21.0
1	0	1	14,883	255	15,225.8	14,883.0	14,883.0
1	0	0	11,371		11,028.2	11,371.0	11,371.0
0	0	1	0	6,170.3	5,662.2	3,497.9	3,582.9
0	0	0	0		508.1	2,672.5	2,113.2

7.2.3   Graphical representation of log-linear models

These relations among collapsible models can be more easily read and interpreted using ideas derived from graph theory (see for example Whittaker [29]). First we make an interaction graph for the log-linear model by placing nodes which correspond with the variables in the model, and then adding edges, lines connecting any pairs of nodes corresponding to variables where an interaction is included in the model. Note that 3-way or higher order interactions give rise to edges for all pairs of variables in the graph, but no attempt is made to represent these higher order interactions. Therefore, more than one model may give rise to the same graph. Where a series of edges connects two nodes, we will say that there is a path between them.

Figure 7.1 shows the interaction graphs for a selection of the models from the Dutch census example. In models where A and B are not connected by any edges, so that there is no path from A to B, the contingency table can be collapsed over all of the covariates in the graph. So in the top row of Figure 7.1, the contingency table $A \times B \times X_1$ can be collapsed over X_1 in model M_1 and in model M_2. This illustrates corollary 1 graphically, that under models M_1 and M_2 the population-size estimate is identical to the population-size estimate under M_0. Similarly, the table $A \times B \times X_1 \times X_2$ can be collapsed over

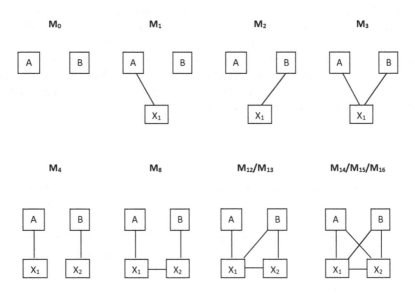

FIGURE 7.1
Interaction graphs for log-linear models taken from the Dutch census example.

both X_1 and X_2 in model M_4, as there is no path from A to B. In passing, we note that this property of model M_4 shows that the inclusion probabilities of A and of B may both be heterogeneous as long as the sources of heterogeneity, i.e., X_1 and X_2, are not related.

To analyse more complex interaction graphs, we need to introduce the notion of a short path. A short path from A to B is a path that does not contain a sub-path from A to B (for full technical details, see [29]). In M_{12}/M_{13} in Figure 7.1 there is a short path $A-X_1-B$, but $A-X_1-X_2-B$ is not a short path, because $A-X_1-B$ is a sub-path of it. In M_8, however, $A-X_1-X_2-B$ is a short path, as it contains no sub-paths. There is one further complication, which is that a short path need not be *short* (that is, have few nodes). In models with more variables, there may be more than one path from A to B without any common nodes (other than A and B), they may have different lengths, and yet both are short paths because neither contains a sub-path from A to B. N_1 in Figure 7.4 shows an example, where $A-X_3-B$ and $A-X_2-X_1-B$ are both short paths.

In models with a short path connecting A and B, the table is not collapsible over the covariates in the path. A simple example is model M_3 of Figure 7.1, where the contingency table $A \times B \times X_1$ cannot be collapsed over X_1. Similarly M_8 cannot be collapsed over either X_1 or X_2 since both are on the short path from A to B. When the covariate X_2 is not part of any path from A to B as in models M_5 and M_9, then $A \times B \times X_1 \times X_2$ is collapsible over X_2.

Models M_{14}, M_{15} and M_{16} all have the same interaction graph, and this graph has two short paths from A to B, $A-X_1-B$ and $A-X_2-B$. Since both of the covariates appear on one of the short paths, the model is not collapsible over either of the covariates. Note that two models which share the same graph must share the same collapsibility, and if they are collapsible, then the two models will be collapsible to the same model. For example, models M_{12} and M_{13} have the same graph, and are collapsible to model M_3.

7.2.4 Three registers

In this section we give illustrative examples of the situation with three or more registers. For three registers A, B and C the contingency table $A \times B \times C$ has one structural zero cell. We consider how the corollaries apply to the context of three registers A, B and C, and with a single covariate X using three models with their graphs displayed in Figure 7.2.

For model $M_{17} = [AX][AB][BC]$ the table $A \times B \times C \times X$ is collapsible over covariate X as it is not on any short path. The other models where A and C are conditionally independent given B, and X is related to only one of the registers (not illustrated), namely models $[AB][BC][BX]$ and $[AB][BC][CX]$ demonstrate corollary 2. For model $M_{18} = [ABX][BCX]$ covariate X is on a short path from A to C and therefore the contingency table is not collapsible over X. (Note that $A-B-C$ is also a short path, but we do not consider short paths that imply collapsing over registers with this theory, since if we

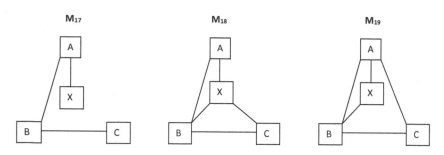

FIGURE 7.2
Interaction graphs for selected simple models in a three-register system with one covariate.

collapse over registers there are not enough data points to estimate the parameters). For model $M_{19} = [ABX][BC][AC]$, covariate X is not on the short path from A to B, as the short path is $A - B$, and therefore the contingency table is collapsible over X. The maximal model $[ABX][BCX][ACX]$ (not illustrated) for the three-register situation with one covariate is discussed at the end of Appendix A of van der Heijden et al. [27], who also discuss graph representations of the models and collapsibility.

The advantage of being able to use more than two registers is that the restrictive (conditional) independence assumption between variables A and B can be replaced by less restrictive assumptions. For example, in the situation of three registers without covariates, the saturated model is the model with all two-factor interactions. Now it is possible to search for more restrictive models that still describe the data well. See, for example, Bishop et al. [4] for details.

7.3   Applications of invariant population-size estimation

7.3.1   Modelling strategies with active and passive covariates

We have already noted that including passive covariates in a model provides a practical approach to breaking down the estimated population by given covariates, even though they have no influence on the estimated population size. However, the introduction of many covariates may lead to sparse contingency tables and hence to numerical problems due to empty marginal cells in those margins that are fitted. Consider, for example, a saturated model such as $[AX_1 X_2 X_3][BX_1 X_2 X_3]$, in which the conditional odds ratios between A and B are 1 at all levels of the covariates. When a zero count for $(A, B) = (1, 1)$

occurs in one of the $X_1 \times X_2 \times X_3$ subtables, the estimate in this subtable for the missing population is infinite. One way to solve this is by removing the higher-order interaction parameters to reduce the sparsity – for example, using $[AX_1][AX_2][AX_3][BX_1][BX_2][BX_3]$ (although such a draconian removal of all the three- and four-factor interactions may not be necessary).

Another approach to tackle this numerical instability problem is to start with an analysis using only active covariates, for example, using the covariates observed in all registers in the saturated model. We may monitor the usefulness of the model by checking the size of the point estimate and its confidence interval. If the usefulness is problematic (for example, when the upper bound of the parametric bootstrap confidence interval is infinite), we may make the model more stable by choosing a more restrictive model. One way to do this is by making a covariate passive. For example, starting again from model $[AX_1X_2X_3][BX_1X_2X_3]$, both in model $[AX_1X_2][BX_1X_2X_3]$ and model $[AX_1X_2X_3][BX_1X_2]$ the covariate X_3 is passive and both models yield identical estimates and confidence intervals. If one of these two models is chosen, the model may then be expanded by adding further passive variables, such as variables that are only observed in register A or register B (see Section 7.4).

7.3.2 Working with invariant population-size estimates

We may consider how the existence of invariant population sizes can help in choosing an appropriate model. First, it is important to note that, if two models yield the same population-size estimate in theory, then they are numerically equally stable, with the same variances. Therefore, efficiency cannot be used as a criterion for choosing between models with invariant population-size estimates.

Second, the criterion that can be used to choose between models with invariant population-size estimates is model fit. If the models are fitted to the same data (that is, with the same covariates and therefore the same table size), we choose the model that has a better fit to the data. For example, in Table 7.1, models M_5, M_9, M_{10}, M_{12} and M_{13} all yield an estimate of the population size of 5,696.1, and they are all fitted to the four-variable table built up by variables A, B, X_1 and X_2. Yet the model of choice is M_{12} as it has the lowest AIC value. So we would like to break down the total population-size estimated for the four-way table using this model, as it provides the best fit to the observed values in the four-way table.

Notice that this estimate of 5,696.1 for the unobserved part of the population is also found for the model M_3 fitted to the three-way table built up by variables A, B and X_1. This illustrates that in the four-variable table the variable X_2 is a passive variable in the models M_5, M_9, M_{10}, M_{12} and M_{13}. Notice that M_3 is a saturated model in the three-variable table, so it has perfect fit. What we learn from the AIC measures of models M_5, M_9, M_{10}, M_{12} and M_{13} is that, if we want to extend model M_3 with the passive variable X_2, then it is best to do this using model M_{12}. This model not only has the lowest AIC value, it also has an adequate fit with a deviance of 4.9 with 6 degrees of freedom.

What if we are happy with a model for the three-variable table like model M_3, and there turns out *not* to be a model where X_2 is a passive variable with an adequate fit for the four-variable table? Even though the fit of such a model is not adequate, the model has the attractive property that the efficiency of this extended model is identical to the efficiency for model M_3. However, the estimates of the breakdown under this extended model will be biased because we include a variable with poor predictive ability which affects the fitted values.

The invariance property can also help with a formal model selection for saturated models (compare with Elliott & Little [9]). For instance, in Table 7.1, there is no statistical test for the choice between models M_0 and M_3, since both are saturated with 0 deviance on 0 df. However, when we eliminate the interaction between A and X_1 or between B and X_1, we obtain models M_1 and M_2 that have identical total population-size estimates to model M_0. For the example in Table 7.1 M_2 has an adequate fit, and model M_1 has an inadequate fit. These results suggest that it is wise not to present estimates for model M_0, as M_0 gives no indication how the estimates should be broken down over values of X_1, whereas the AIC measures of models M_1 and M_2 show that only the breakdown of M_2 is adequate. If models M_1 and M_2 were both to have inadequate fits, then model M_3 should be preferred over model M_0, as it is saturated and has perfect fit.

7.4   Dealing with partially observed covariates

A second challenge from the linkage of registers for population-size estimation is that missing values will be generated by the linkage. The values of variables that are only available in a subset of the registers are not known for the records that are not present in this subset. In the following sections, we present and explore a framework for both estimating undercoverage and completing these missing values in one procedure.

7.4.1   Framework for population-size estimation with partially observed covariates

Figure 7.3 is a graphical representation of the linkage of two registers. The representation shows the linked data with observations (often individuals) in the rows and variables in the columns. The data for variables available only in register A are on the left and denoted by a, the data for variables only in register B are on the right and denoted by b, and in the middle are the data

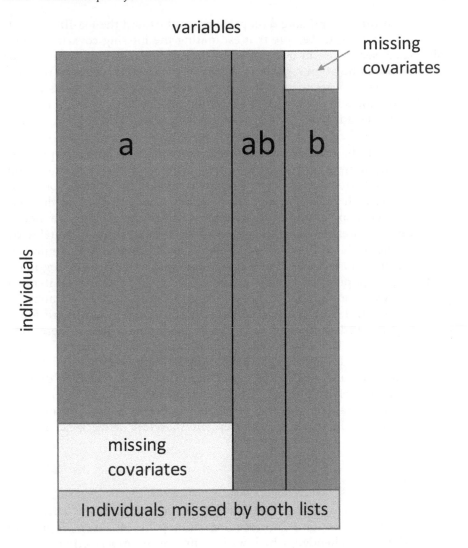

FIGURE 7.3
Graphical representation of two linked registers, see text for details.

for variables that registers *A* and *B* have in common, denoted by *ab*. Typically *ab* includes the variables used to link the registers.

Therefore each register has some unique variables, and these give rise to missing data in the observed cases, shown in Figure 7.3, as well as the requirement to provide breakdowns for the completely missed cases as discussed above.

We treat this as a missing data problem, and extend the log-linear modelling to deal with it. We note that estimating the missing covariate values only makes sense for covariates that pertain to all registers involved. For example, if a population register is linked with a hospital register, then the hospital register covariate type of medical problem should not be estimated for all individuals in the population register who do not appear in the hospital register, as it is likely that they do not have a medical problem at all.

As before we consider categorical covariates only, and allow the log-linear model to be extended to include the covariates. This again provides the flexibility to have independence or conditional independence given the covariates, and widens the range of situations where the assumptions from Section 7.1 (which are still needed for the estimation of the size of the completely missed population) are reasonable. It also allows us to use models which can give breakdowns of this estimate by the categories of suitable covariates. The problem is solved by applying the Expectation-Maximization (EM) algorithm to estimate the missing observations within the log-linear modelling for population-size estimation. The EM algorithm is an iterative procedure where each iteration has an expectation (E) and a maximization (M) step. In the E-step the expectations of the missing values are found given the observed values and the fitted values under a model, here some log-linear model. The E-step yields completed data. Then, in the M-step, the log-linear model is fitted to the completed data and this updates the fitted values that are used in the next E-step (for a more detailed description, see van der Heijden et al. [26]). This proceeds until convergence. The algorithm has linear convergence, which may make the algorithm very slow. Yet the likelihood increases in each step and therefore convergence is guaranteed.

7.4.2   Example

We extend the example from Section 7.2.2 using two different covariates: X_3 represents marital status, observed only in register A, and X_4 represents region where apprehended, which we interpret as an imperfectly measured residence variable, observed only in register B. The resulting four-way contingency table is shown in Table 7.3.

The most complicated (maximal) model that can be fitted to this table is $[AX_4][BX_3][X_3X_4]$, which has eight free parameters – an intercept, four main effects, and three interactions. This is the same as the number of counts in Table 7.3, so it is saturated. Notice that the term AX_3 is not included in the maximal model, because when $A = 0$, X_3 is missing and unknown. Therefore only three counts (two cells and one margin) are available for the term AX_3, which is not enough information to identify it. Note too that violations of this model, such as dependence between A and B conditional on the covariates, cannot be tested, because it is already saturated, so the additional interaction cannot be fitted. However, it is possible to investigate the sensitiv-

TABLE 7.3

People aged 15-64 with Afghan, Iranian or Iraqi nationality from a quality evaluation of the Dutch 2011 Census (see van der Heijden et al. [27]). Co-variate X_3 (marital status) is only observed in population register A and X_4 (police region where apprehended) is only observed in police register B.

		$B = 1$		$B = 0$
		$X_4 = 0$	$X_4 = 1$	X_4 missing
$A = 1$	$X_3 = 0$	259	539	13,898
	$X_3 = 1$	110	177	12,356
$A = 0$	X_3 missing	91	164	-

ity of the outputs to violations of this assumption, see Section 7.5.2. It is also possible to investigate whether more restrictive models also fit the data. For example, we can investigate whether one of the interactions, such as X_3X_4, can be eliminated without the fit deteriorating. If so, the model becomes $[AX_4][BX_3]$, and under this model A and B are (unconditionally) independent, rather than independent conditional on X_3 and X_4. Such a conditional independence assumption is strong, yet in many datasets there may not be enough power to test its correctness. A direct relation between the passive covariates of register A and those in B can only be assessed among those individuals that are in both registers. If the overlap between registers A and B is relatively large, the relationship can easily be assessed; conversely, if the overlap is small, there is little power to establish whether or not this relation should be included in the model.

The maximal model can easily be extended when there are additional covariates. For example, consider the situation that in addition to X_3 being observed in A and X_4 in B, a variable X_5 is observed in A and B; then the maximal and saturated model is $[AX_4X_5][BX_3X_5][X_3X_4X_5]$.

Most registers have several covariates that are not common to other registers, because the different registers are set up with different purposes in mind. In these cases, we can extend the modelling strategy of Section 7.3.1:

- first, determine a small number of covariates of interest, i.e., covariates that are related to being included or not in one or both of the registers. This may require some subject matter knowledge;

- second, undertake a model fitting process with these covariates, where in the final model some of them may turn out to be active, others passive; and

- third, extend the model with (possibly several) passive covariates, which may or may not be register specific.

The passive covariates are helpful in breaking down the population size under the assumption that the passive covariates of register A are independent of the passive covariates of register B conditional on the active covariates. This approach can be particularly useful in those situations where each of the registers has unique covariates, but a description of the full population in terms of all of these covariates is needed.

7.4.3 Interaction graphs for models with incomplete covariates

The models for tables with incomplete covariates may be analysed using interaction graphs as before, but care must be taken that a postulated graph can be fitted with the available data – since the reduced number of cells means that some interactions cannot be fitted, as we have already seen. Models with the same diagram as M_{12}/M_{13} from Figure 7.1 (but with X_1 and X_2 replaced by X_3 and X_4, respectively), for example could not be fitted from Table 7.3 as the BX_3 term cannot be estimated.

Figure 7.4 shows the interaction graphs for a selection of possible models with two and three covariates. In model I_5 there is only one short path, and both covariates are on it, so the model cannot be collapsed over any covariates. In I_1 there are two short paths $A - X_3 - B$, and another $A - X_2 - X_1 - B$ (remember that a short path is not necessarily short, but does not contain a subpath); since all the covariates appear on at least one short path, the model again cannot be collapsed over any covariates. I_2 is a reduced (not maximal) model which can be collapsed over both X_2 and X_1.

7.4.4 Results of model fitting

Table 7.4 shows the fitted values from fitting the saturated $[AX_4][BX_3][X_3X_4]$ model to the data from Table 7.3. The counts 13,898 and 12,356, and the counts 91 and 164, have been distributed over the levels of the missing variables. For example, 13,898 is distributed over the levels of X_2 into 4,510.8 and 9,387.2, and the ratio of these two counts is equal to the ratio of the observed counts 259 and 539. As a result, in Table 7.4 the odds ratio for the counts 259, 539, 110 and 177 (that is, the top-left corner subtable) is projected to the subtables on the right and at the bottom for which at most the margins were available in Table 7.3.

Van der Heijden, Zwane and Hessen [28] and van der Heijden et al. [27] consider models for this example with five covariates. X_1, gender, X_2, age (four levels), and X_5, nationality (1 = Iraqi; 2 = Afghan; 3 = Iranian) are all fully observed in the two registers; as before, X_3, marital status, is observed only in the municipal register (GBA), and X_4, police region where apprehended (1 = in one of the four largest cities of the Netherlands, 2 =

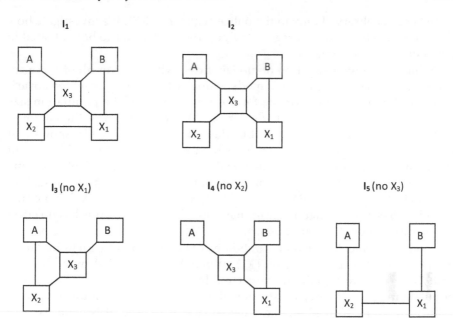

FIGURE 7.4
Interaction graphs for examples of models for the two-covariate and three-covariate tables described in Section 7.4.2.

TABLE 7.4
People aged 15-64 with Afghan, Iranian or Iraqi nationality from a quality evaluation of the Dutch 2011 Census (see van der Heijden et al. [27]), completed with the fitted values from the $[AX_4][BX_3][X_3X_4]$ model.

		$B = 1$		$B = 0$	
		$X_2 = 0$	$X_2 = 1$	$X_2 = 0$	$X_2 = 1$
$A = 1$	$X_1 = 0$	259.0	539.0	4,510.8	9,387.2
	$X_1 = 1$	110.0	177.0	4,735.8	7,620.3
$A = 0$	$X_1 = 0$	63.9	123.5	1,112.4	2,150.2
	$X_1 = 1$	27.1	40.5	1,167.9	1,745.4

elsewhere), is observed only in the police register (HKS). We investigate how these models allow the coverage of the population register to be evaluated in terms of the five covariates.

Table 7.5 shows a selection of models fitted with some or all of these covariates. Model N_1 is a saturated model using only the fully observed covariates. For this model the estimate for the missed part of the population size is 5,504.6, and the total population size is 33,098.6. However, the parametric bootstrap confidence interval (Buckland and Garthwaite [5]) shows that this solution is numerically unstable, as the upper bound of the 95% percent confidence interval is infinite. The instability is a consequence of too many active covariates, and one solution is to make covariate X_5 passive, such as in models $N_2 = [AX_1X_2][BX_1X_2X_5]$ and $N_3 = [AX_1X_2X_5][BX_1X_2]$. Note that model N_2 has a somewhat better fit than model N_3, but that this has no effect on the estimated total population.

Models N_2 and N_3 are both candidates to be extended by including marital status (X_3) or police region (X_4). When N_2 is extended by adding X_3 and X_4 as passive variables, we get model N_4, which yields an identical population-size estimate, illustrating that X_3 and X_4 are indeed passive. With 72 degrees of freedom and a deviance of 75.7, the fit is good. We can check whether it is better to make covariates X_3 and X_4 active and we do this by adding the interaction between them to give model N_5. The deviance of this model is unchanged but uses an extra df, and its AIC is therefore higher, and we conclude that N_4 is a better working model than N_5. We also extend N_3 by adding X_3 and X_4 as passive variables giving N_6. The population-size estimate is identical to N_3 as expected, but the deviance is much larger than for N_4 and N_5 (we cannot compare with the deviance of N_3 since they derive from different tables). Adding the interaction between X_3 and X_4 gives N_7 which improves the deviance, but it is still larger than the deviance of N_4, which we therefore choose as the final model.

7.5 Precision and sensitivity

There are two estimation problems (Figure 7.3): estimating the missing covariates and estimating the number of individuals (and their covariate values) missed by both registers. For both estimation problems, we are interested in the precision when the model assumptions are true, and the sensitivity of the outcomes to deviations from the model assumptions.

7.5.1 Precision

The precision (here an overall term referring to the variance of the estimates) of the estimates for the missing covariates depends on the precision of the

TABLE 7.5
Models fitted to examples of variables A, B and X_1 to X_5, with their deviances, degrees of freedom, AICs, estimated population size and 95% confidence intervals.

Model	Deviance	df	AIC	Pop.size	CI
models fitted to the $A \times B \times X_1 \times X_2 \times X_5$ (sub)table					
N_1 $[AX_1X_2X_5]\|[BX_1X_2X_5]$	0	0	144.0	33,098.6	32,209 - ∞
N_2 $[AX_1X_2]\|[BX_1, X_2X_5]$	24.9	16	136.8	33,504.1	32,480 - 35,468
N_3 $[AX_1X_2X_5]\|[BX_1X_2]$	28.8	16	140.7	33,504.1	32,480 - 35,468
models fitted to the $A \times B \times X_1 \times X_2 \times X_3 \times X_4 \times X_5$ table					
N_4 $[AX_1X_2X_4]\|[BX_1X_2X_3X_5]$	75.7	72	315.7	33,504.1	32,480 - 35,468
N_5 $[AX_1X_2X_4]\|[BX_1X_2X_3X_5]\|[X_3X_4]$	75.7	71	317.7	33,503.8	32,395 - 35,543
N_6 $[AX_1X_2X_4X_5]\|[BX_1X_2X_3]$	523.8	72	763.7	33,504.1	32,480 - 35,468
N_7 $[AX_1X_2X_4X_5]\|[BX_1X_2X_3]\|[X_3X_4]$	289.1	71	531.4	33,510.9	32,363 - 35,432

estimate of the odds ratio under the model. This precision is directly related to the size of the population that is in both A and B: the larger this size, the smaller the standard error of the odds ratio and the standard errors of the estimates of the missing covariates, and the better the precision.

The precision of the estimates for the individuals missed by A and B is the outcome of two sources: first, the precision of the estimates of the missing covariates that we just discussed, and, second, the implied coverage. Precision of the estimates of the missing covariates has a direct impact on the precision of the data for the individuals missed. The second source of imprecision is related to the implied coverage. From Table 7.3, the population coverage of register A implied by B is $259/(259 + 63.9) = 0.802$. However, the coverage of police register B implied by A is only $259/(259 + 4,510.8) = 0.057$. If either or both of these implied coverages are low, the estimated number of missed individuals is large relative to the number of individuals seen, and hence imprecise.

Estimates of the precision can be obtained using the parametric bootstrap (Buckland and Garthwaite [5]). The parametric bootstrap provides a simple way to find the confidence intervals when the contingency table is not fully observed. To compute the bootstrapped confidence intervals for a specific log-linear model, first compute the population size under this model and therefore the multinomial probabilities of the completed data (a probability in each cell of the table, summing to 1). Draw a multinomial sample with the completed population size with these probabilities, and then aggregate it to the same form as the observed data (for example, a sample drawn from Table 7.4 is aggregated to the form of Table 7.3, including inserting the structural zeroes). The specific log-linear model used is then fitted to the resulting data, giving an estimate of the population size. This process is repeated L times. By ordering the L bootstrapped population-size estimates, a percentile confidence interval can be constructed.

7.5.2 Sensitivity

We are also interested in the sensitivity of the estimates to violations of the assumptions of the model. It is clearly possible to investigate whether maximal models can be reduced by setting some parameters equal to zero. However, it is not possible to test directly whether interaction parameters that are not included in the maximal model should be included, as there is not enough data to estimate them. Gerritse et al. [12] show that for a particular dataset we can investigate how sensitive the outcome of the maximal model is to the assumption that certain parameters are zero. For example in model M_5, the three two-factor interactions AX_1, BX_2 and AB and all three- and four-factor interactions are zero. Assume that we are interested in investigating to what extent the population-size estimate is dependent on the assumption that the AX_1 interaction is zero. Extend the maximal model with a *fixed* parameter value for the AX_1 interaction, $\lambda^{AX_1} = k$. The model may be fitted as a log-linear Poisson regression with offset $\exp(\lambda^{AX_1})$. This model can be fitted for a range of values of k, and appropriate values can be chosen by making use of the fact that log-linear parameters are closely related to odds ratios.

Gerritse et al. [14] argue that the sensitivity of outcomes of the analyses to violation of the independence assumption and to violation of perfect linkage is larger when the implied coverage is lower. In the absence of covariates, this can be explained as follows. Let m_{ij} be the expected count for cell (i, j) $(i, j = 0, 1)$, where m_{00} is the missing count to be estimated. Under independence, the odds ratio is 1, i.e., $m_{00}m_{11}/m_{01}m_{10} = 1$, so that $m_{00} = m_{01}m_{10}/m_{11}$. Under dependence with odds ratio θ, $m_{00}\theta = m_{01}m_{10}/m_{11}$. Thus, the smaller the overlap in cell $(1,1)$, and hence the smaller the coverage, the larger the estimated value for cell $(0,0)$, and this holds both for independence and dependence. In the same way, when links are missed, this increases the expected values m_{01} and m_{10} and decreases m_{11}, with the result that m_{00} is larger, and this effect is larger the smaller the overlap m_{11}.

7.5.3 Comparison of the EM algorithm with the classical model

Van der Heijden et al. [26] carried out simulations to compare the behaviour of the classical model (denoted by standard loglinear model (LL)), where incomplete covariates are ignored, with the model where incomplete covariates are completed with the EM algorithm (denoted by EM). The reference data is generated from a population with four variables, A, B, X_1 and X_2, with prespecified marginal probabilities. Assuming independence, this leads to 16 probabilities, and using these a model is fitted where prespecified conditional odds ratios are used as offsets (see Section 7.5.2). The fitted values of this model are used as the probabilities for the true model, giving marginal probabilities identical to the marginal probabilities of the independence model and conditional odds ratios equal to the prespecified conditional odds ratios.

The simulations, using a variety of different scenarios, show that when the conditional independence assumptions are fulfilled, the EM approach does better, though sometimes only slightly better, than LL. When the conditional independence assumptions are violated, the bias can be substantial in both approaches, in particular when the inclusion probabilities (or equivalently the implied coverages) in one or more sources are low.

7.6 An application when the same variable is measured differently in both registers

We may apply these methods in the situation that two registers both measure the same variable, but where the measure in one register is considered to be more reliable, or valid, than the other. This is closely related to the classical two-phase sampling problem, where there is an inexpensive but low quality measurement which can be obtained from a large sample, and a more expensive and more accurate approach which is used on a subsample. Two-phase sampling concentrates on combining the small sampling variance of the large sample measure with the measurement accuracy of the small sample measure. As an alternative, we apply the EM algorithm to complete the missing information on the highest quality measure, and additionally to provide this information for statistical units which are missed in both registers (a situation which cannot generally be handled by two-phase sampling).

7.6.1   Example: Injuries in road accidents in the Netherlands

The example we use is the number of serious injuries in road accidents in the Netherlands. In the Netherlands, the number of serious injuries is an indicator used in assessing the road safety target. There are two sources of information on serious road injuries, namely the police and hospitals, both usually involved after such an accident. The police are supposed to record the accident and its cause in the police crash record database, but this regularly does not happen for some reason, such as that it is not clear which police officer has to file the accident report, or that the injury is not considered very serious. The hospital that treats the seriously injured can report the cause of the injury in the hospital inpatient registry but this is sometimes forgotten and then such a patient's connection to a traffic accident is lost. Thus both registers have coverage problems; for more details, see Reurings and Stipdonk [22].

For the year 2000, Reurings and Stipdonk [22] present the data in the upper panel of Table 7.6. The police register has a larger undercoverage than the hospital register. It is reasonable to assume that, where the police registers do record the mode of transport of injured persons, they do this more accurately than the hospital, because assessing the cause of accidents is a more important function for the police, where liability plays a role, than for the hospital, which is more concerned about the health-related issues than on the cause and details of the accident. So we treat the Cause of accident variables as different variables in the two registers even though they measure the same concept. Notice that in the fully observed 2×2 subtable, there are 287 joint classifications not in agreement where the police recorded the involvement of a motorized vehicle but the hospital recorded that no motorized vehicle was involved, and 29 vice versa.

One approach is to define a new variable Y with three levels conflating X_2 and B, namely $(X_2 = 1, B = 1), (X_2 = 2, B = 1)$ and $(X_2 = \text{missing})$, and then fit the model $[AX_1][X_1 Y]$ with X_1 values missing for $A = 0$ (see Panel 2 of Table 7.6). Alternatively, we can use the EM algorithm approach, assuming that the hospital Cause of accident is missing for those accidents only registered by the police, and that the police Cause of accident is missing for those accidents

TABLE 7.6
Road accidents in the Netherlands in 2000, from Reurings and Stipdonk [22].
Motorized vehicle involved X_1 is only observed in the police register (A) and
motorized vehicle involved X_2 is only observed in the hospital register (B).
Levels of X_1 and X_2 are 1 = yes, 2 = no.

Panel 1: Observed counts

		$B = 1$		$B = 0$	
		$X_2 = 1$	$X_2 = 2$	X_2 missing	Total
$A = 1$	$X_1 = 1$	5,970	287	1,351	7,608
	$X_1 = 2$	28	256	70	354
$A = 0$	X_1 missing	2,947	4,120	-	7,067
Total		8,945	4,663	1,421	15,029

Panel 2: Fitted values under $[AX_1][X_1Y]$

		$B = 1$		$B = 0$	
		$X_2 = 1$	$X_2 = 2$	X_2 missing	Total
$A = 1$	$X_1 = 1$	5,970.0	287.0	1,351.0	7,608.0
	$X_1 = 2$	28.0	256.0	70.0	354,0
$A = 0$	$X_1 = 1$	2,509.6	120.6	567.9	3,198.1
	$X_1 = 2$	437.4	3,999.4	1,093.6	5,530.4
Total		8,945.0	4,663.0	3,082.5	16,690.5

Panel 3: Fitted values under $[AX_2][BX_1][X_1X_2]$

		$B = 1$		$B = 0$		
		$X_2 = 1$	$X_2 = 2$	$X_2 = 1$	$X_2 = 2$	Total
$A = 1$	$X_1 = 1$	5,970.0	287.0	1,289.0	62.0	7,608.0
	$X_1 = 2$	28.0	256.0	6.9	63.1	354,0
$A = 0$	$X_1 = 1$	2,933.2	2,177.6	633.3	470.2	6.214,3
	$X_1 = 2$	13.8	1,942.4	3.4	478.8	2.438,4
Total		8,945.0	4,663.0	1.932,6	1,074.1	16,614.7

only registered by the hospital, and then fitting model $[AX_2][BX_1][X_1X_2]$ (see Panel 3 of Table 7.6). The estimates from the two models are quite different – 1661.5 accidents missed by both sources are estimated by the first approach, and 1585.7 by the second, and the split over the levels of X_1 is very different. Notice that the odds ratio in the observed $A = 1$ and $B = 1$ subtable is very large (in the upper part of Table 7.6 it is almost 200), and in both approaches this odds ratio is used to find the estimates in the subtables of $A = 0$ and $B = 1$.

We prefer the EM algorithm approach using model $[AX_2][BX_1][X_1X_2]$ to the model $[AX_1][X_1Y]$ (which is equivalent to the approach of Reurings and Stipdonk [22]). This cannot be based on model fit, as both models are saturated and have a perfect fit. Instead, we make a professional judgement; it is reasonable to assume that, for example, the count 2,947 for which X_1 is missing, should be split over motorized and non-motorized in the same way as when X_1 is not missing. Additional support comes from the model-based bootstrap which gives smaller confidence intervals for estimates based on $[AX_2][BX_1][X_1X_2]$. This is in line with Elliott and Little's [9] principle 5 for choosing between saturated models, using the model that gives estimates with the smallest variance.

7.6.2   More detailed breakdown of transport mode in accidents

A more detailed coding of motorized mode of transport into six categories was used for data from 2010 (Reurings and Bos [21]):

1. Sitting in a car

2. Riding a motorbike

3. Riding a moped

4. Bicycles in a motorized accident

5. Pedestrians in a motorized accident

6. Other in a motorized accident

This finer coding into 7 levels is potentially useful for assessing the cause of a rise or decline in accidents. The observed data (Table 7.7) on this classification have only occasional, low off-diagonal counts, and this is attractively plausible – we do not want "non-motorized" to be frequently mixed up with "sitting in a car". Some of the properties of the 2010 data are different too. The police registered many fewer accidents than in 2000, and at the same time the coverage of the hospital register increased.

Using log-linear model $[AX_1][X_1Y]$ with Y extended to eight categories gives unstable results; iterative fitting produces two lines of estimates where $A = 0$ consist of 0s only, which means that the model never converges. Therefore only results for the approach using model $[AX_2][BX_1][X_1X_2]$ are shown in Table 7.8. The stability of the estimates was investigated using the parametric bootstrap (see Section 7.5.1), giving the results in Table 7.9. The seven estimated total numbers of severely injured are rather stable. We conclude that the refined classification of "motorized" into 6 categories in 2010 is usable for policy purpose when model $[AX_2][BX_1][X_1X_2]$ is applied.

7.7 Discussion

In this chapter we have presented a methodological framework that is useful for modelling and analysis of statistics based on linked registers where additional categorical auxiliary variables are available. The methodology has potential for simultaneously solving the problems of undercoverage and of missing covariate values for those cases missed in some or all of the registers. This corresponds to solving the missing data problem for the missing covariates and missed individuals parts in Figure 7.3. The EM algorithm can also be used to solve the problem of missing data in covariates that are incompletely measured. If there is only a single register, this is a simple missing data problem, but in the case of multiple registers the extra information can be used to provide breakdowns of estimates by these variables. The software we employ, the CAT-procedure in R, is able to handle this problem (see Meng and Rubin [18]; Schafer [23] and [24] for more details). For three registers the problem of incomplete covariates has been studied by Zwane and van der Heijden [34], who show that for this problem the EM algorithm can easily be adapted.

The simulation studies show that, in comparison with the classical method where those partially observed covariates are ignored, the EM approach performs slightly better when the underlying MAR assumption and the conditional independence assumption for inclusion in the registers is met. When these assumptions are violated, both models can be severely biased.

Theoretically, the methodology can also be used when the number of covariates is large, where stability can be improved by making some of the covariates passive (see van der Heijden et al. [27]). In this instance there is little practical experience and we hope that this methodology will be used more so that the practical benefits become clearer.

7.7.1 Alternative approaches

Multiple imputation provides an alternative method for dealing with missing values in covariates. It was used, next to EM, by Gerritse et al. [13]. They argue that multiple imputation is more flexible, and in Table 7.4 where the persons in the two cells labelled missing would be expected to be most simi-

TABLE 7.7

Road accidents in the Netherlands in 2010. Data from Reurings and Bos (p.25) [21]. Motorized vehicle involved X_1 is only observed in police register (A) and Motorized vehicle involved X_2 is only observed in hospital register (B). m.a. = motorized accident.

Observed counts

X_1	$B=1$ X_2							$B=0$ X_2	Total
	1	2	3	4	5	6	7	missing	
$A=1$									
1. Sitting in car	856	7	12	26	61	62	18	130	1,172
2. Driving motorbike	3	261	33	0	7	5	2	20	331
3. Driving moped	7	83	504	19	8	60	21	47	749
4. Bicycles in m.a.	55	2	10	523	38	29	139	96	892
5. Pedestrians in m.a.	9	0	2	11	208	33	3	35	301
6. Other in m.a.	20	1	18	4	7	17	2	22	91
7. Non-motorized	2	0	0	9	1	7	82	12	113
$A=0$									
missing	1,100	860	1,530	844	482	540	8,578	-	13,934
Total	2,052	1,214	2,109	1,436	812	753	8,845	362	17,583

TABLE 7.8

Motorized vehicle involved X_1 is only observed in Police register (A) and Motorized vehicle involved X_2 is only observed in hospital register (B). Fitted values (rounded) under model $[AX_2][BX_1][X_1X_2]$.

| | $B = 1$ | | | | | | | | $B = 0$ | | | | | | | |
| | X_2 | | | | | | | | X_2 | | | | | | | |
X_1	1	2	3	4	5	6	7	Total	1	2	3	4	5	6	7	Total
$A = 1$																
1. Sitting in car	856.0	7.0	12.0	26.0	61.0	62.0	18.0	1,042.0	106.8	0.9	1.5	3.2	7.6	7.7	2.2	129.9
2. Driving motorbike	3.0	261.0	33.0	0.0	7.0	5.0	2.0	311.0	0.2	16.8	2.1	0.0	0.5	0.3	0.1	20.0
3. Driving moped	7.0	83.0	504.0	19.0	8.0	60.0	21.0	702.0	0.5	5.6	33.7	1.3	0.5	4.0	1.4	47.0
4. Bicycles in m.a.	55.0	2.0	10.0	523.0	38.0	29.0	139.0	796.0	6.6	0.2	1.2	63.1	4.6	3.5	16.8	96.0
5. Pedestrians in m.a.	9.0	0.0	2.0	11.0	208.0	33.0	3.0	266.0	1.2	0.0	0.3	1.4	27.4	4.3	0.4	35.0
6. Other in m.a.	20.0	1.0	18.0	4.0	7.0	17.0	2.0	69.0	6.4	0.3	5.7	1.3	2.2	5.4	0.6	21.9
7. Non-motorized	2.0	0.0	0.0	9.0	1.0	7.0	82.0	101.0	0.2	0.0	0.0	1.1	0.1	0.8	9.7	11.9
$A = 0$																
1. Sitting in car	989.1	17.0	31.7	37.1	89.1	157.2	578.3	1,899.5	123.4	2.1	4.0	4.6	11.1	19.6	72.1	236.9
2. Driving motorbike	3.5	634.1	87.2	0.0	10.2	12.7	64.3	812.0	0.2	40.8	5.6	0.0	0.7	0.8	4.1	52.2
3. Driving moped	8.1	201.6	1,331.8	27.1	11.7	152.1	674.7	2,407.1	0.5	13.5	89.2	1.8	0.8	10.2	45.2	161.2
4. Bicycles in m.a.	63.6	4.9	26.4	745.6	55.5	73.5	4,465.7	5,435.2	7.7	0.6	3.2	89.9	6.7	8.9	538.6	655.6
5. Pedestrians in m.a.	10.4	0.0	5.3	15.7	303.8	83.7	96.4	515.3	1.4	0.0	0.7	2.1	40.0	11.0	12.7	67.9
6. Other in m.a.	23.1	2.4	47.6	5.7	10.2	43.1	64.3	196.4	7.4	0.8	15.2	1.8	3.3	13.7	20.5	62.7
7. Non-motorized	2.3	0.0	0.0	12.8	1.5	17.7	2,634.4	2,668.7	0.3	0.0	0.0	1.5	0.2	2.1	313.0	317.1

TABLE 7.9

Parametric bootstrap point estimates of causes according to the police, with 95 percent confidence interval (percentile method) and median, under model $[AX_2][BX_1][X_1X_2]$

	mean	2.5 percent	median	97.5 percent
1. Sitting in car	3,307.1	2,997.9	3,300.6	3,644.9
2. Driving motorbike	1,195.0	1,071.2	1,190.8	1,336.8
3. Driving moped	3,317.2	2,996.6	3,312.6	3,657.6
4. Bicycles in m.a.	6,981.8	6,382.4	6,980.5	7,583.1
5. Pedestrians in m.a.	884.7	749.4	881.0	1.046.1
6. Other in m.a.	350.8	238.4	344.3	498.9
7. Non-motorized	3,099.3	2,565.1	3.094.4	3,646.8

lar to persons not in the population register, imputing from this subpopulation is easily accomplished using multiple imputation. But this approach is separate from the estimation of the unobserved part of the population, and it does not benefit from the integrated way of dealing with these two issues.

Multiple imputation is however more natural in the case of continuous covariates, as used in Zwane and van der Heijden [33]. Further research into the benefits of improving estimation using continuous covariates is also desirable. A more general strategy for modelling and estimation with linked registers which includes options for using categorical and continuous auxiliary variables could then emerge. An important element of that would be to have more examples of the usefulness of the approaches described in this chapter.

An alternative modelling approach is based on the Rasch model, which allows for heterogeneity of inclusion probabilities for different sources (i.e., some individuals have higher probabilities and others lower), under a more restrictive range of assumptions about the model structure. Such models can also be fitted to strata defined by a breakdown of records by a particular variable, in which case estimates of the breakdown of the completely missed cell will result; these estimates do not necessarily share the dependence structure of the observed part because of the effect of the latent variables which capture the heterogeneity of inclusion probabilities. This approach can be parameterised in a simple way as there is a corresponding log-linear model for each Rasch model. For more details see Pelle et al. [19].

7.7.2 Quality issues

The approaches presented here deal only with the problem of undercoverage. However, many registers also contain overcoverage, and this can have an effect on the undercoverage estimation by increasing the number of records to be linked. This will generally inflate population-size estimates by inflating the number of records appearing in only one register, though it could have the opposite effect if the overcovered records appear in both registers. A way to estimate overcoverage was developed for the Estonian Census (Lehto et al. [17]). The general idea is to accept people who are registered only as residents if they show up in other administrative sources and show, as such, a sign of life. Zhang [30,31] also provides a framework for models to deal with overcoverage error, but for both approaches it is important to have at least one source that does not suffer from overcoverage in order to make a suitable adjustment. Zhang and Dunne [32] present the trimmed dual-system estimator as a way to estimate the population size with overcoverage. More work is needed on how the estimation of undercoverage and overcoverage can be integrated into a coherent set of procedures.

In situations with two sources for the same concept, a practical approach is often taken where a new, composite variable is created. The values of the variable from register A are used when they are available, the values from register B are used for observations missed by register A, and some ad hoc solution (such as an imputation procedure) is used for observations that were missed by both registers. In the approach presented here, however, for those observations that were missed by register A we translate the values in register B into estimates of what would have been found in register A using the observed $A = 1$ and $B = 1$ subtable to give the required structure. This has the potential to form part of a principled approach to the analysis of linked data as called for in Smith and Chambers [25], because it uses the full data to make an estimate of the required concept, rather than taking the data at face value whether it meets the required concept or not.

Bibliography

[1] ASMUSSEN, S. and EDWARDS, D. (1983). Collapsibility and response variables in contingency tables. *Biometrika* **70** 567–578.

[2] BAKER, S.G. (1990). A simple EM algorithm for capture-recapture data with categorical covariates (with discussion). *Biometrics* **46** 1193–1197.

[3] BARTOLUCCI, F. and FORCINA, A. (2001). Analysis of capture-recapture data with a Rasch-type model allowing for conditional dependence and multidimensionality. *Biometrics* **57** 714–719.

[4] BISHOP, Y.M.M., FIENBERG, S.E., and HOLLAND, P.W. (1975). *Discrete Multivariate Analysis*. Cambridge: M.I.T. Press.

[5] BUCKLAND, S. and GARTHWAITE, P. (1991). Quantifying precision of mark–recapture estimates using the bootstrap and related methods. *Biometrics* **47** 255–268.

[6] CHAO, A., TSAY, P.K., LIN, S.-H., SHAU, W.-Y. and CHAO, D.-Y. (2001). The applications of capture-recapture models to epidemiological data. *Statistics in Medicine* **20** 3123–3157.

[7] CORMACK, R.M. (1989). Log-linear models for capture-recapture. *Biometrics* **45** 395–413.

[8] DI CECCO, D. (2019). Estimating population size in multiple record systems with uncertainty of state identification. In Zhang, L.-C. & Chambers, R.L. (eds) *Analysis of Integrated Data*. Chapter 8, 169–196. Boca Raton: Chapman & Hall/CRC.

[9] ELLIOTT, M.R. and LITTLE, R.J.A. (2000). A Bayesian approach to combining information from a census, a coverage measurement survey, and demographic analysis. *Journal of the American Statistical Association* **95** 351–362.

[10] FIENBERG, S.E. (1972). The multiple recapture census for closed populations and incomplete 2^k contingency tables. *Biometrika* **59** 591–603.

[11] FIENBERG, S.E., JOHNSON, M.S. and JUNKER, B.W. (1999). Classical multilevel and Bayesian approaches to population size estimation using multiple lists. *Journal of the Royal Statistical Society, Series A* **162** 383–405.

[12] GERRITSE, S.C., VAN DER HEIJDEN, P.G.M. and BAKKER, B.F.M. (2015). Sensitivity of population size estimation for violating parametric assumptions in log-linear models. *Journal of Official Statistics* **31** 357–379.

[13] GERRITSE, S.C., BAKKER, B.F.M. and VAN DER HEIJDEN, P.G.M. (2015). Different methods to complete datasets used for capture-recapture estimation, *Statistical Journal of the IAOS* **31**, 613–627.

[14] GERRITSE, S.C., BAKKER, B.F.M. and VAN DER HEIJDEN, P.G.M. (2016). *The impact of linkage errors and erroneous captures on the population size estimator due to implied coverage.* Den Haag/Heerlen: Statistics Netherlands, discussion paper 2017-16.

[15] IWGDMF - INTERNATIONAL WORKING GROUP FOR DISEASE MONITORING AND FORECASTING (1995). Capture-recapture and multiple-record systems estimation I: History and theoretical development. *American Journal of Epidemiology* **142** 1047–1058.

[16] KIM, S.-H. and KIM, S.-H. (2006). A note on collapsibility in DAG models of contingency tables. *Scandinavian Journal of Statistics* **33** 575–590.

[17] LEHTO, K., MAASING, E. and TIIT, E.-M. (2016). Determining permanent residency status using registers in Estonia. European Conference on Quality in Official Statistics (Q2016) Madrid, 31 May–3 June 2016.

[18] MENG, X.L. and RUBIN, D.B. (1991). IPF for contingency tables with missing data via the ECM Algorithm. In *Proceedings of the Statistical Computing Section of the American Statistical Association* 244–247. Washington D.C.: American Statistical Association.

[19] PELLE, E., HESSEN, D.J. and VAN DER HEIJDEN, P.G.M. (2016). A log-linear multidimensional Rasch model for capture–recapture. *Statistics in Medicine* 35 622–634.

[20] POLLOCK, K.H. (2002). The use of auxiliary variables in capture-recapture modelling: an overview. *Journal of Applied Statistics* 29 85–102.

[21] REURINGS, M.C.B. and Bos, N.M. (2012). *Ernstig Verkeersgewonden in de Periode 2009 en 2010. Update van de Cijfers* (R-2012-7). Leidschendam: Stichting Wetenschappelijk Onderzoek Verkeersveiligheid SWOV. Available at: www.narcis.nl/publication/RecordID/oai:library.swov.nl:129380.

[22] REURINGS, M.C. and STIPDONK, H.L. (2011). Estimating the number of serious road injuries in the Netherlands. *Annals of Epidemiology* 21 648–653.

[23] SCHAFER, J. (1997). *Analysis of Incomplete Multivariate Data*. Boca Raton: Chapman & Hall/CRC.

[24] SCHAFER, J. (1997). *Imputation of missing covariates under a general linear mixed model*. Pennsylvania: PennState University, Department of Statistics.

[25] SMITH, P.A. and CHAMBERS, R.L. (2018). Discussion of "Statistical challenges of administrative and transaction data" by David J. Hand. *Journal of the Royal Statistical Society, Series A* 181 585.

[26] VAN DER HEIJDEN, P.G.M., SMITH, P.A., CRUYFF, M. and BAKKER, B. (2018). An overview of population size estimation where linking registers results in incomplete covariates, with an application to mode of transport of serious road casualties. *Journal of Official Statistics* 34 239–263.

[27] VAN DER HEIJDEN, P.G.M., WHITTAKER, J., CRUYFF, M., BAKKER, B. and VAN DER VLIET, R. (2012). People born in the Middle East but residing in the Netherlands: Invariant population size estimates and the role of active and passive covariates. *Annals of Applied Statistics* 6 831–852.

[28] VAN DER HEIJDEN, P.G.M., ZWANE, E. and HESSEN, D. (2009). Structurally missing data problems in multiple list capture-recapture data. *Advances in Statistical Analysis* **93** 5–21.

[29] WHITTAKER, J. (1990). *Graphical Models in Applied Multivariate Statistics*. New York: John Wiley & Sons.

[30] ZHANG, L.-C. (2015). On modelling register coverage errors. *Journal of Official Statistics* **31** 381–396.

[31] ZHANG, L.-C. (2019). Log-linear models of erroneous list data. In Zhang, L.-C. & Chambers, R.L. (eds) *Analysis of Integrated Data*. Chapter 9, 197–218. Boca Raton: Chapman & Hall/CRC.

[32] ZHANG, L.-C. and DUNNE, J. (2018). Trimmed dual system estimation. In Böhning, D., van der Heijden, P.G.M. & Bunge, J. (eds.) *Capture-recapture methods for the social and medical sciences* 229–235. Boca Raton: CRC Press.

[33] ZWANE, E. and VAN DER HEIJDEN, P. (2008). Capture-recapture studies with incomplete mixed categorical and continuous covariates. *Journal of Data Science* **6** 557–572.

[34] ZWANE, E.N. and VAN DER HEIJDEN, P.G.M. (2007). Analysing capture-recapture data when some variables of heterogeneous catchability are not collected or asked in all registrations. *Statistics in Medicine* **26** 1069–1089.

[35] ZWANE, E.N., VAN DER PAL-DE BRUIN, K. and VAN DER HEIJDEN, P.G.M. (2004). The multiple-record systems estimator when registrations refer to different but overlapping populations. *Statistics in Medicine* **23** 2267–2281.

8

Estimating population size in multiple record systems with uncertainty of state identification

Davide Di Cecco

Istat, National Institute of Statistics, Rome, Italy

CONTENTS

8.1 Introduction

We consider the problem of estimating the size of a population of interest, or "target population", by integrating multiple data sources. Each source provides a list of the units of our population. In this context, we identify three possible scenarios:

1. Each unit of our target population is included in at least one of the sources, but the identification of the units is not error free: Some

out-of-scope units are erroneously included in the lists and, vice versa, some units of our population are erroneously identified as out-of-scope;

2. All observed units are correctly identified as belonging or not to the target population. However, some units are not enlisted in any of the available sources. So, we have a problem of undercoverage of our lists;

3. Not all units are comprised in the data at hand, and the observed units are not correctly classified with respect to the target population.

The first scenario can be essentially characterized as a case of misclassification. We can exploit the information redundancy at our disposal to estimate the misclassification errors by making some assumptions on the randomness of that redundancy, and, as a result, we could even estimate unit-level probabilities of belonging to the target population.

The second scenario represents a typical situation of a capture–recapture setting, where we have a set of lists which are incomplete (they do not cover all units, and some unobserved units are not registered in any list) and overlapping (a unit can be registered in several sources). The event of being captured corresponds to the event of being registered in a list. Unlike the previous scenario, we can just estimate the number of unobserved units.

In the third scenario, which is the focus of this chapter, we are assuming that both issues, of uncertainty of detection and uncertainty of state identification, are present in the data at hand. We essentially refer to a capture–recapture setting where the classic assumption of absence of error in the units identification is relaxed. In this context, a misclassification error can be rephrased as an "erroneous capture".

Note that, in the capture–recapture setting we are considering, captures are not controlled by the researchers. Capture occasions, in fact, consist of a collection of reporting systems usually set up for different purposes. Broadly speaking, the fact of being covered by a source reflects an event of administrative nature.

There are several possible reasons for an erroneous capture. In general, the scopes of the various sources may differ to a great extent and may differ with our target population. So, each list may contain different subpopulations of out–of–scope units. Differences in scope may be due to not easily detectable differences in the definitions of the units. For example, the available information pertaining to the registered events, their temporal description, their legal definition, may vary in each source, and the harmonization of these aspects may not be error free, and, consequently, the assignment of the units to our target population may not be error free. There may be errors due to delays in registration/cancellation in each list. When merging the sources, errors in identifying the units can cause duplicates or erroneous linkages.

Obviously, any piece of available information should be included in the process of identification of the erroneous cases in our lists, and, ideally, recognizing and deleting spurious cases should constitute a first phase of our analysis, after which some capture–recapture technique could be utilized on the "clean" data.

However, in many cases, the available information does not suffice to single out every false capture, and there will remain a certain portion of uncertainty for which we have no capability of discerning the cause of error.

Our approach for these cases consists in defining all possible (residual) errors as random classification errors, and we propose an unsupervised approach based on the use of a latent class model to estimate them. The reference to a classification error in this context is easily acceptable whenever the belonging to the target population is defined by the value of an auxiliary variable. For example, if units are defined as persons in a certain age range, the uncertainty of the variable "age" will be entirely reflected in the classification of the units. However, this view of the problem is useful even when such a classification mechanism is not evident per se.

The use of latent variables to model the "true" latent status of a unit is common practice in behavioral and social sciences. In these areas, the latent variable usually represents a concept which is not directly observable and can only be measured through related indicators. A different conception, described in-depth in [4], is when the observed (manifest) variables are actual measurements of the desired variable, and the latent variable, representing its true value, is well-defined and measurable, with a known number of modalities. In this context, we consider each measurement as possibly affected by error, and the redundancy of information allows us to estimate the latent variable and to evaluate the accuracy of each source of measurement in an unsupervised way. A clear example of application of this conception is in latent class models for multiple diagnostic test (see [38] for an overview), where each test has a false positive and a false negative rate, and the latent variables identify the true medical status. In our model we refer to this latter conception, where the binary latent variable identifies the in–scope and out–of–scope units, and the manifest variables, i.e., the observed captures, can be either false or true, and can be interpreted as potentially erroneous measurements of that latent variable. Note that both the observed and the latent variables are binary. There are examples of capture–recapture models with a different number of modalities for the observed and the latent variables, namely in multistate models (see, e.g., [27]), to account for uncertainty in state assignment. However, we will limit ourselves to the binary case.

In Section 8.2 we describe the model in formal terms and give several details on estimation, identifiability and model selection. In Section 8.3 we describe the use of covariates to model heterogeneity of capture probabilities, and a particular case of uncatchable subpopulations. In Section 8.4 we provide some observations on the tenability of the use of the latent class we propose in our model. In Section 8.5 we present a Bayesian approach to the model.

8.2   A latent class model for capture–recapture

Formally, let k be the number of lists (or capture occasions). Let Y_i be the random variable indicating whether a unit has been captured by the i–th source (i.e., is listed on the i–th list):

$$Y_i = \begin{cases} 1 & \text{if the unit is captured in the } i\text{–th list;} \\ 0 & \text{otherwise.} \end{cases}$$

We have an array of k variables $\underline{Y} = (Y_1,\ldots,Y_k)$ representing the captures of a unit in k different lists. The observed binary array for each unit (often called capture history in capture–recapture literature) is denoted as $\mathbf{y} = (y_1,\ldots,y_k)$.

Let U be the set of all units that can be captured by any list, i.e., the union of the target populations of the k sources. Let U_1 be our target population. It has to be $U_1 \subset U$, so that there are not any uncatchable units. We can manage a situation with units that are uncatchable for some sources, but not all (see Section 8.3.2). The cardinality of U, $|U|$ is denoted as N, while $|U_1| = N_1$.

The observed units can be classified in a 2^k contingency table $T = [n_{\underline{y}}]_{\underline{y} \in \{0,1\}^k}$ where $n_{\underline{y}}$ denotes the number of units having capture history $\underline{y}$. The total number of observed units is n_{obs}, while the units having capture history $\underline{y} = \underline{0} = (0,\ldots,0)$ are unobserved, so that $\sum_{\underline{y} \neq \underline{0}} n_{\underline{y}} = n_{obs}$, and $N = n_{obs} + n_{\underline{0}}$.

The latent variable identifying the units belonging to our target population is denoted as X:

$$X = \begin{cases} 1 & \text{if a unit belongs to } U_1; \\ 0 & \text{otherwise.} \end{cases}$$

While $n_{x,\underline{y}}$ denotes the number of units belonging to latent class x which present capture history $\underline{y}$, so that $\sum_{x \in 0,1} n_{x,\underline{y}} = n_{\underline{y}}$. Note that $n_{1,\underline{0}}$ is the number of units in U_1 that are not captured, while $n_{0,\underline{0}}$ is the number of uncaptured units which are in U but not in U_1.

For the model distribution we will use the notation introduced in [14], for which $P(\underline{Y} = \underline{y}) = \pi_{\underline{y}}$. Superscripts will eventually denote the random variables when they are not clear from the context so that, for example, $\pi_{10}^{XY_k}$ denotes the probability $P(X = 1, Y_k = 0)$. The class of models we are considering can be expressed as the following mixture model:

$$\pi_{\underline{y}} = \sum_{x \in 0,1} \pi_x \pi_{\underline{y}|x}, \tag{8.1}$$

where $\pi_{\underline{y}|x}$ is the conditional probability $P(\underline{Y} = \underline{y} \mid X = x)$, and π_x is the marginal probability of belonging to the latent class x, i.e., the probability of a random unit to belonging or not to the target population, and constitutes the weight parameter of the mixture model. As a consequence, the likelihood function is:

$$L(\pi_{\underline{y}} ; n_{\underline{y}}) \propto \prod_{\underline{y}} \pi_{\underline{y}}^{n_{\underline{y}}} = \prod_{\underline{y}} \left(\sum_x \pi_x \pi_{\underline{y}|x} \right)^{n_{\underline{y}}}.$$

The simplest model we can consider in this class is the following:

$$\pi_{\underline{y}} = \sum_{x \in 0,1} \pi_x \prod_{i=1}^{k} \pi_{y_i|x}, \tag{8.2}$$

which satisfies the local independence assumption, i.e., the manifest variables are independent conditionally on X.

Applications of this class of models in capture–recapture to model unobserved heterogeneity in capture probabilities are known (see, e.g., [1]). Note, however, that the local independence assumption is hardly tenable in our setting. As we have said, we are considering a setting of captures which are not controlled by the researcher. Hence, even if we condition on the value of the latent variable, there would likely be both positive and negative dependencies between the captures of a same unit in different lists, as a unit recorded in one list may be more (or less) likely to appear on a second one than a unit that has not been recorded on that list.

Various methods have been proposed to model dependencies in latent class models, for example random effects ([28], [8]), or multiple latent variables ([10]). In a Bayesian context, [23] proposed a Dirichlet process prior to model dependencies in a non-parametric way. However, the classic approach dating back to [11] in capture–recapture is to directly model these dependencies via loglinear models; hence, in order to explore more complex dependence structures between the manifest and latent variables, we are going to consider a class of hierarchical loglinear models with a latent variable X which interacts with all the manifest variables.

It has been demonstrated (see, e.g., [15]) that model (8.2) is equivalent to the loglinear model

$$[XY_1]\dots[XY_k], \tag{8.3}$$

where we use the classic notation of hierarchical loglinear models reporting only the higher order interaction terms (sometimes called generators). Model (8.3) is the simplest case we will consider: Each additional interaction parameter with respect to (8.3) represents a deviation from the local independence assumption. The term Local Dependence model is sometimes used for this setting. For an application of this model to capture–recapture see, e.g., [5] and [31].

Our goal is to estimate the dimension of our target population, N_1, i.e., we want to estimate the number of units (both captured and uncaptured) for which $X = 1$. We have $N_1 = \sum_{\underline{y} \neq \underline{0}} n_{1,\underline{y}} + n_{1,\underline{0}}$, that is, N_1 equals the sum of the number of in–scope units for each capture history, including the units not captured in any list.

We note in passing that a different, more cautious, utilization of our model would limit the interest to the estimates of the parameters. We could consider the estimated parameters to evaluate the coverage error rates, either at a source-level, or at an individual-level. At a source-level, for each source Y, we can estimate the following four conditional probabilities: $\begin{pmatrix} \pi_{0|0}^{Y|X} & \pi_{1|0}^{Y|X} \\ \pi_{0|1}^{Y|X} & \pi_{1|1}^{Y|X} \end{pmatrix}$. The extradiagonal elements represent the two possible error rates of that source: $\pi_{0|1}^{Y|X}$ is the undercoverage rate of Y, while $\pi_{1|0}^{Y|X}$ constitutes its over-coverage or false capture rates. $\pi_{1|1}^{Y|X}$ is the probability of a unit in U_1 of being captured by source Y. The general propensity of a random unit of being captured by source Y is

$$\pi_1^Y = \pi_{1|0}^{Y|X} \pi_0^X + \pi_{1|1}^{Y|X} \pi_1^X.$$

At an individual-level, we can consider the so-called posterior probabilities $\hat{\pi}_{x|\underline{y}_j}$, where $\underline{y}_j$ is the capture history of unit j, which estimates the probability of that unit to belong to class x. Instead of a direct estimation of N_1, we can utilize those posterior probabilities in a separate capture–recapture analysis. In fact, once we have individual probabilities of being out–of–scope, they can be easily integrated in various ways in a capture–recapture estimator to account for false captures. Typically, those probabilities are estimated by means of an ad hoc survey and a supervised model, see, for example, [13], [17], [22]. In our model, we do not necessarily need an error-free audit sample to estimate them.

8.2.1 Decomposable models

Here we present a sub-class of models, namely the decomposable models, which will be useful in the following. To define that class, we first introduce the dependency graph. The dependency graph G of a joint categorical distribution is the undirected simple graph where nodes represent variables, and any two nodes are adjacent unless the relative variables are independent (conditionally on all other variables). A model is said to be graphical if its dependency graph G exhaustively defines the structure of its joint distribution, i.e., if the joint distribution has no constraint other than those defined by G. Note that, in a graphical loglinear model, this is equivalent to say that the model includes all and only the interaction terms represented by the cliques of G, i.e., the generators of the model are the cliques of G. A model is said to be decomposable if its dependency graph G is chordal, i.e., if any cycle of G of four or more nodes has a chord (an edge that is not a part of the cycle connecting two of its vertices). If a model is decomposable, it can be showed ([9]) that the joint distribution factorizes in a product of conditional distributions. In detail, let $\{\mathcal{C}_1, \dots, \mathcal{C}_g\}$ be the maximal cliques of the associated graph G, $\left(\bigcup_{i=1}^g \mathcal{C}_i = G \right)$. Then, there exists an ordering $\{\mathcal{C}_{\sigma(1)}, \dots, \mathcal{C}_{\sigma(g)}\}$ of the set, such that, if we define the set of separators $(S_2,\ldots,S_g)$ as

$$S_i = C_{\sigma(i)} \cap \bigcup_{j=1}^{i-1} C_{\sigma(j)} \qquad i = 2,\ldots,g,$$

we have that

- each separator S_i is a clique of G;
- each S_i is contained in one of the maximal cliques $C_{\sigma(j)}$, $\sigma(j) < i$;
- the joint distribution can be written in the following form:

$$\prod_{i=1}^{g} \pi^{C_i} \left(\prod_{j=2}^{g} \pi^{S_j} \right)^{-1}, \tag{8.4}$$

where π over a (sub)graph is to be intended as the (marginal) distribution over the variables included in the (sub)graph.

Then, if we denote the conditional probabilities π^{C_i}/π^{S_i} as $\pi^{C_i|S_i}$ (with a slight abuse of notation), we can write the decomposable distribution as the following product:

$$\pi^{C_1} \prod_{i=2}^{g} \pi^{C_i|S_i}.$$

In our models the latent variable X interacts with all other variables, so X is included in all maximal cliques and in all separators; then, π^{C_1} can be written as $\pi^X \pi^{C_1|X}$. As a consequence, we can always write a decomposable distribution as:

$$\pi^X \pi^{C_1|X} \prod_{i=2}^{g} \pi^{C_i|S_i}. \tag{8.5}$$

As an example, refer to the graph with four sources $\underline{Y} = (A, B, C, D)$ of Figure 8.1.

We have just two hierarchical loglinear models with that dependence graph:

$$[AX][BX][CX][DX][CD] \quad \text{and} \quad [AX][BX][CDX].$$

Just the second one is graphical and, since the graph is chordal, it is decomposable. In fact, we have $C_1 = \{A, X\}$, $C_2 = \{B, X\}$, $C_3 = \{C, D, X\}$, $S_2 = \{X\}$, $S_3 = \{X\}$, and the model can be written as:

$$\pi_{\underline{y}} = \sum_x \pi_x \pi_{a|x} \pi_{b|x} \pi_{cd|x}. \tag{8.6}$$

Decomposable models have some computational advantages, as we will see throughout the remainder of this chapter.

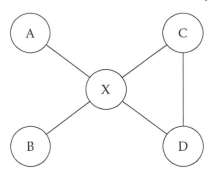

FIGURE 8.1
A simple decomposable graph with 5 nodes (4 sources and a latent variable).

8.2.2 Identifiability

An obvious necessary condition for the class of models we are analyzing to be identifiable is that the number of independent parameters is not larger than the number of distinct observed profiles $\mathbf{y}$ minus one. For this reason, a latent class model with 3 or less manifest binary variables and a structural zero is intrinsically non-identifiable. In fact, in that case, we can observe at most 7 distinct profiles, and the simplest possible model (8.3) would have 7 parameters. In order to use such a model, we can either impose some restrictions on the parameters, or add some covariates as manifest variables to the model.

To our knowledge, a general necessary and sufficient condition for identifiability of this class of models has not been found. Results for sub-class of models can be found in [32] and [2]. There exists, however, a necessary and sufficient condition for local identifiability: Goodman [14] defined a model as locally identifiable if the maximum likelihood solution is unique within some small interval (i.e., an ϵ–neighborhood) around the solution. This is less restrictive than global identifiability, in which the MLE must be unique throughout the entire parameter space. Goodman showed that a necessary and sufficient condition for local identifiability is that the Jacobian of the expected response patterns with respect to the parameters is of full column rank. Most computer programs for latent class analysis include a check of this condition.

8.2.3 EM algorithm

To compute the maximum likelihood estimate (MLE) of our models, we assume the existence of a complete contingency table $T^* = [n_{x,\mathbf{y}}]$, of which we observed the marginal counts T. We define a loglinear model for T^* with parameters $\{\lambda\}$, then, once we initialize an estimate of the parameters $\{\lambda\}$ (or, alternatively, of the cells of T^*), we can set an EM algorithm iterating over the following two steps:

E-step: compute the expected counts of the cells of T^* conditionally on the observed marginal T and the current estimate of $\{\lambda\}$ by calculating, for each profile $\underline{\mathbf{y}}$, the estimated posterior probability $\hat{\pi}_{x|\underline{\mathbf{y}}}$ and

$$\hat{n}_{x\underline{\mathbf{y}}} = n_{\underline{\mathbf{y}}}\hat{\pi}_{x|\underline{\mathbf{y}}};$$

M-step: update the MLE of the parameters $\{\lambda\}$ of the loglinear model over the frequencies $\hat{n}_{x\underline{\mathbf{y}}}$ computed in the E-step.

In the M-step, one can use the Iterative Proportional Fitting (IPF) algorithm to obtain the MLE of the $\{\lambda\}$. Actually, it has been demonstrated that the IPF does not need to fully converge: a single iteration at each M-step should suffice for the whole EM algorithm to converge (see, e.g., [30]).

It is in theory possible to adopt two approaches to the estimation process: the conditional and the unconditional one (see [11]). In the conditional approach, we would maximize the loglikelihood conditionally on the structural zero cells which would be excluded from both steps in the algorithm. Once the EM algorithm converges, the current MLEs of $\{\lambda\}$ are used to estimate each cell of T^* including the structural zero cells $n_{x,0}$. On the other hand, the unconditional approach would amount to estimate and update the structural zero cell n_0 at each iteration. To do this, we should use a "nested" EM algorithm (see, e.g., [6]), with two nested cycles, where the outer cycle initializes and updates an estimate of the structural zero cell $\hat{n}_0$, while the inner one updates the estimate of the complete data T^* (including cells $\hat{n}_{x,0}$), based on the current $T \cup \hat{n}_0$. However, since the two approaches are equivalent, the first is preferable as it is computationally much easier.

Whenever the model is decomposable, we can avoid the loglinear expression, and the M-step of the algorithm can be solved in a closed form, as we have an analytic formula for the parameters maximizing the complete loglikelihood. That is, since the complete loglikelihood can be expressed as

$$\pi_{x,\underline{\mathbf{y}}} = \pi_{\mathcal{C}_1} \prod_{i=2}^{g} \pi_{\mathcal{C}_i|\mathcal{S}_i},$$

we can estimate each parameter $\pi_{\mathcal{C}_i|\mathcal{S}_i}$ with $\hat{n}_{\mathcal{C}_i}\big/\hat{n}_{\mathcal{S}_i}$ where $\hat{n}_{\mathcal{C}_i}$ and $\hat{n}_{\mathcal{S}_i}$ are obtained by summing the $\hat{n}_{x,\underline{\mathbf{y}}}$ resulting from step E over the variables not included in the two subgraphs.

As an example, in estimating model (8.6), the M-step would amount to calculate:

$$\hat{\pi}_x = \frac{\hat{n}_x}{n_{obs}}, \qquad \hat{\pi}_{a|x} = \frac{\hat{n}_{xa}}{\hat{n}_x}, \qquad \hat{\pi}_{b|x} = \frac{\hat{n}_{xb}}{\hat{n}_x}, \qquad \hat{\pi}_{cd|x} = \frac{\hat{n}_{xcd}}{\hat{n}_x},$$

where

$$\hat{n}_x = \sum_{y \neq \underline{0}} \hat{n}_{x\underline{y}}, \quad \hat{n}_{xa} = \sum_{b,c,d} \hat{n}_{xabcd}, \quad \hat{n}_{xb} = \sum_{a,c,d} \hat{n}_{xabcd}, \quad \hat{n}_{xcd} = \sum_{a,b} \hat{n}_{xabcd}.$$

8.2.4 Fixing parameters

The unsupervised approach of the latent class models we are adopting implies a neutral perspective on the quality (i.e., the evaluation of the error rates) of each source. For this reason, latent class models are sometimes used to estimate the error rates in a set of sources when there is no available audit sample. However, if we have some reliable information on the error rates of the sources, we can include it in the model by fixing some parameters. For example, if we know that source Y has no false captures, we can fix the corresponding parameter $\pi_{1|0}^{Y|X}$ equal to zero.

Another reason to impose some restrictions on parameters is to save degrees of freedom by assuming a more parsimonious model. In some cases this can be necessary in order to have an identifiable model. That is, whenever the number of available lists does not permit the use of more complex models.

As long as we consider decomposable models, and the kind of restrictions we want to impose is fixing a parameter to a specific value or imposing an equality constraint between parameters, we can maximize the constrained likelihood with a rather simple procedure.

Goodman in 1974 proposed a simple procedure which works in most cases. Mooijaart and van der Heijden [24] then improved Goodman's algorithm, and proposed to estimate a (decomposable) model with equality and/or fixed-value restrictions on the probabilities by adding Lagrange multipliers to the log-likelihood function to be maximized.

8.2.5 A mixture of different components

We want to point out a hypothesis underlying the class of loglinear models we are considering, which limits the relations we can represent: namely, the fact that the two component distributions (conditioned on $X = 1$ and $X = 0$) have the same dependence structure. This aspect can be unsuitable in some situations.

For example, under model $[AX][BX]$, A and B will be conditionally independent whether X equals 1 or 0. If we add term $[AB]$, we would allow dependencies between A and B conditionally on X, but the parameter of interaction $[AB]$ would be the same whether $X = 1$ or $X = 0$. If we add term $[ABX]$ too, then the interaction between A and B will not necessarily be the same conditionally on $X = 1$ and $X = 0$, but such a strategy of including higher order interactions would likely lead to overparameterization. It is not difficult to imagine a situation where A and B are independent in just one of the sub-populations, so that, say,

$$\pi_{ab|0} = \pi_{a|0} \, \pi_{b|0}, \quad \text{but} \quad \pi_{ab|1} \neq \pi_{a|1} \, \pi_{b|1}.$$

For example, say we have two sources with the same question identifying a binary status. Say a "0" status is irrelevant for the respondent, while a "1" identifies a sensitive status which is likely to be held hidden. In this case, the conditional independence hypothesis may hold under $X = 0$, but when $X = 1$, the respondent may tend to be consistent in lying in both surveys.

In such a case, we would like to fix the constraint $\pi_{ab0} = \pi_{a|0}\pi_{b|0}$ in model [ABX], but this is not the kind of constraints which are easily handled in the estimating algorithm (see Section 8.2.4). A possible workaround is to abandon the loglinear form, and define a mixture model of two unequal component distributions:

$$\pi_{ab} = \pi_0 \pi_{a|0} \pi_{b|0} + \pi_1 \pi_{ab|1}.$$

In this way, the two components can be defined independently (eventually as two distinct loglinear models). This generalization poses no difficulties to the estimation process. In fact, in step M of the algorithm, we can treat independently the part of the complete loglikelihood pertaining to the values $\hat{n}_{1,\mathbf{y}}$ estimated in step E, and the part pertaining to values $\hat{n}_{0,\mathbf{y}}$.

An example of usage of this kind of models can be found in mixture models for record linkage, where [18] proposed a model where the loglinear component relative to the true matches has main effects only, while the second component has all main effects and two-way interactions. That is, the model satisfies the conditional independence assumption just under $X = 1$. The idea is that, if we have a true match, the disagreements on the various manifest variables are the result of random errors, while in a non-match, it is more likely that disagreements occur in logical sets ([42]).

8.2.6 Model selection

Model selection is a crucial point in capture–recapture, as often the estimates of the population size are very sensitive to small changes in the choice of the parameters.

A typical approach to the problem is to use information criteria to compare models. In particular, one can use a stepwise (backward or forward) procedure for parameter selection based on AIC or BIC. However, we have several results in the literature specifically designed for latent class models. Most contributions in this respect focus on detecting pairwise dependencies between manifest variables when using local independence model (8.2).

A first approach is that of comparing the estimated and observed bivariate marginals $\pi^{Y'Y''}$ and $\hat{\pi}^{Y'Y''}$ for any pair of sources Y' and Y''.

Qu et al. [28] propose to inspect the estimated correlations

$$\widehat{Corr}_{Y'Y''} = \frac{\hat{\pi}_{11}^{Y'Y''} - \hat{\pi}_1^{Y'}\hat{\pi}_1^{Y''}}{\sqrt{\hat{\pi}_1^{Y'}(1-\hat{\pi}_1^{Y'})\,\hat{\pi}_1^{Y''}(1-\hat{\pi}_1^{Y''})}},$$

and the observed one, and calculate the differences for each pair of variables. Significant differences can be spotted with the eventual use of bootstrap confidence intervals, and the corresponding pairwise interaction term can be added to the loglinear model.

Bivariate Residuals (BVR) compute the Chi-squared distance between the two bivariate distributions:

$$\sum_{i,j\in\{0,1\}^2} \frac{\left(n_{i,j}^{Y'Y''} - \hat{n}_{i,j}^{Y'Y''}\right)^2}{\hat{n}_{i,j}^{Y'Y''}}.$$

Apparently (see [25]), the BVR index does not have an asymptotic χ^2 distribution as stated in [40]. Then, again a bootstrap confidence interval can be employed.

Another statistic which received some attention in recent years is the Expected Parameter Change (EPC), known as Modification Index in Structural Equation Modeling, which allows one to detect which parameters not included in a model are more relevant ([25], [26]). We consider a null model defined by parameters θ_0, and an alternative model defined by parameters $\theta = (\theta_0, \theta_1)$. The null model is seen as a model where parameters θ_1 are restricted to be equal to zero. The EPC then estimates the shift in the parameters θ_1 if they were freed. Let $\mathbf{s}_{\theta_1}$ be the score vector of the fixed parameters, $\mathbf{s}_{\theta_1} = \partial l(\theta)/\partial\theta_1$, and let $\hat{I}(\theta)$ be the expected information matrix. Then, if the null model holds, the test statistic

$$\mathbf{s}_{\theta_1}^T \hat{I}(\theta)^{-1}\mathbf{s}_{\theta_1},$$

evaluated at $\theta_1 = 0$ is asymptotically distributed as a χ^2 with degrees of freedom equal to the number of parameters in θ_1.

That being said, the best model selected by any of the above tools not always is the one giving the best estimates of N_1, and it may be the case that different models with a similar level of goodness-of-fit lead to very different estimates of the population size.

Then, it is important to incorporate in model assessment some considerations on the subject matter. The estimates of the parameters should be evaluated on the basis of our prior knowledge of the data, and models with unreasonable values for certain parameters should be discarded.

Note that the parameters resulting from the algorithm of Section 8.2.3 refer to the truncated distribution conditionally on the structural zero cells. Hence, to have a clear interpretation of the parameters we first have to estimate the complete contingency table T^*, and derive the parameters of the complete distribution.

8.3 Observed heterogeneity of capture probabilities

8.3.1 Use of covariates

Even if we model all the existing dependencies between the lists, there can still be some unexplained variability due to differences in the individual propensity to be captured. Some of this heterogeneity of the capture probabilities can be explained by a set of covariates which have to be included in the model.

Failing to model either list dependencies or heterogeneity in individual capture probabilities usually leads to biased estimates of the population size. Note, however, that the two aspects are not easily disentangled. A loglinear model for $\mathbf{y}$ with the sufficient number of parameters can represent any distribution on $\{0,1\}^k$; so, it is in principle always possible to find a (possibly overparameterized) model with an arbitrary level of goodness-of-fit. Failing to include a relevant source of individual heterogeneity in the model can induce spurious dependencies between the lists, and a spurious parameter can be erroneously included in the model.

The use of individual covariates is usually included in the model in one of two ways (for an analysis of the two approaches in loglinear models, see [37]). A first approach consists in using the observed covariates to affect the model parameters. That is, we reparameterize the probabilities of membership in the latent class (the mixture weights π_x) and/or the conditional capture probabilities $\pi_{\mathbf{y}|x}$ in terms of a set of covariates. In this approach, the covariates explicitly have the role of explanatory variables; they can be both categorical or continuous, and it is possible to limit their influence on specific parameters. The reparameterization can be done in several ways by exploiting the formulation of a loglinear model in terms of multinomial logit models (see [39], [43], [35], [36]).

The second approach, available only if all covariates are categorical, is that of including them as manifest variables. That is, we define a model for the joint distribution $\pi_{\mathbf{y},\mathbf{v}}$ where $\underline{\mathbf{V}}$ is our set of covariates. In this way, we can model the relations between all manifest variables (either captures or covariates) and tune their dependence graph in several ways.

It has to be noted that, in this approach, a covariate does not necessarily impact the estimate of N_1, unless its dependencies with the other manifest variables are carefully chosen. The invariance of the estimate is related to the collapsibility of loglinear models. Say we have a loglinear model M over a set of variables $\underline{\mathbf{Y}}$. We consider a subset of variables $\underline{\mathbf{Y}}'$, and the relative induced loglinear model M' which maintains all and only the parameters of M relative to variables in $\underline{\mathbf{Y}}'$. M is said to be collapsible over $\underline{\mathbf{Y}}'$ if the estimate of each cell $n_{\mathbf{y}'}$ by model M' equal the corresponding marginal estimate by model M. Let G be the dependency graph of M, and H the induced sub-graph over $\underline{Y}/\underline{Y}'$, i.e., the graph of the variables of G not included in $\underline{Y}'$. Asmussen and Edwards [3] found a necessary and sufficient condition for collapsibility, that is, M is collapsible over $\underline{Y}'$ iff the boundary in G of every connected component of H constitutes a generator of M'. Hence, if a loglinear model for $\pi_{x,\underline{y},\underline{v}}$ is collapsible over $(X, \underline{Y})$, then the covariates do not alter the estimate of N_1. On this aspect see also Chapter 7 of this book.

As an example, say we want to include a covariate V in the model of Figure 8.1. Then, model

$$[AX][BX][CDX][CDV]$$

is collapsible over $\{A, B, C, D, X\}$, as C and D are connected in the graph, and V would not alter the final estimate of N_1. On the converse, model

$$[AX][BX][CDX][AV][CV]$$

is not collapsible, as A and C are not connected, and V would influence the estimate of N_1.

Note, however, that this is valid just for the observed variables: if V interacts with the latent X, it will not, in general, be collapsible. For example, model

$$[AVX][BX][CDX],$$

in which the overcoverage error of list A varies by V is not collapsible over $\{A, B, C, D, X\}$, even if $\{A, X\}$ is a clique of the graph.

It is worth noting that, if we are interested in estimating the population size for each stratum identified by a covariate, then even a collapsible model could be useful, as we would simply estimate the quantities $n_{\underline{0},\underline{v}}^{\underline{Y},\underline{V}}$ for all values $\underline{v}$.

8.3.2 Incomplete lists

An extreme case of heterogeneity in capture probabilities is represented by uncatchable sub-populations, i.e., units having null probability of being captured. Here we treat the case of identified subpopulations, that is, units which are known to be uncatchable for one or more lists (but not all). This problem occurs often in practice, as it takes place whenever a source targets a subset of our population of interest. We do not want to lose the information available in those "incomplete sources", as we will call them; however, we need to consider their incompleteness in the model. That is, if we ignore the structural incompleteness of the lists, and treat the uncatchable units as sampling zeros, the resulting estimates of the population size can be severely biased.

We are assuming a perfect knowledge of the sub-populations (or strata) which cannot be captured by each incomplete list. So, U is partitioned into strata where different sets of incomplete lists do not operate. This can be formalized with a stratifying variable S which identifies the various strata partitioning U. Since S is known, one could adopt the naive approach of estimating separately the sizes of the sub-populations for each stratum with distinct models. Each model would comprise just the lists operating on the relative stratum, and then summing the estimates would result in the total population-size estimate. This approach, however, would considerably limit the range of identifiable models we can consider, as we frequently encounter situations where the incomplete lists have separate targets and there is but a small subset where all lists operate. As an example, consider the simple case of 4 lists and 2 strata, one where all lists operates, and one where just 3 lists operate. In the second strata, we can consider just a few simple loglinear models, and a latent class model would not be identifiable.

An alternative approach is that of treating the unobservable captures of the units not covered by the incomplete lists as missing information. Then, the units in those strata are considered as partially classified, i.e., as if the capture history is partially missing, and we estimate the missing part. Of course, the underlying hypothesis is that, conditionally on the other lists, the distribution of each incomplete list is the same across strata, that is, we are hypothesizing a missing at random (MAR) mechanism (see, e.g., [19], chap. 3.2 in [39]). In case this assumption is tenable, we can include all records in a single model even in presence of incomplete lists. On the other hand, when the assumption is not true, the resulting estimates can be biased.

The use of loglinear models to impute partial missing information in contingency tables is well known (see, e.g., [30]). The use of this approach in capture recapture has been explored in [33] and [34] and, in different terms, in [44]. The extension to the case where a latent variable is present presents no difficulties, as it requires little modifications to the algorithm described in Section 8.2.3. Note, however, that the two aspects (of the latent variable X, and of the incomplete lists) cannot be easily addressed separately with loglinear models. In fact, since X interacts with all manifest variables, a loglinear model for $(X, \underline{Y}, S)$ will not be collapsible over $(\underline{Y}, S)$. So, if someone wanted to preliminarily treat the missingness due to the incomplete lists and then estimate the latent class model on the complete data, they should define and estimate two distinct loglinear models.

Say that the stratifying variable S takes values in a finite set $\mathcal{S} = \{s_1, s_2, \ldots\}$ identifying the different strata where different sets of incomplete lists do not operate. For each stratum s we observed a different set of marginal counts T_s. We assume the existence of a complete contingency table $T^* = \left[n_{x, \underline{y}, s}\right]$, and we want to estimate

$$N_1 = \sum_{\substack{\underline{y} \in \{0,1\}^k \\ s \in \mathcal{S}}} n_{1, \underline{y}, s}.$$

Then, in the E-step of the EM algorithm, for each stratum s, we compute the expected counts $\hat{n}_{x, \underline{y}, s}$ of the corresponding cells of T^* conditionally on the observed counts T_s.

Attention should be posed to the choice of the dependence structure of the model for T^*. In fact, in many cases, S, or some of its modalities, can constitute a useful covariate to explain some of the heterogeneity in the capture probabilities, and we would like to include some interaction parameters of S with other variables, in such a way that the model is not collapsible over $(X, \underline{Y})$. However, the MAR assumption of the incomplete lists holds if and only if the probability of being in a certain stratum is independent from all lists that do not operate in that stratum. Hence, a model where S interacts with the incomplete lists variables would violate the MAR assumption and would not be identifiable without further hypothesis. On the converse, a model where S does not interact with any other variable implies the missingness mechanism to be completely at random (MCAR).

In presence of incomplete lists, we have more than one structural zero cell to estimate. Their number varies in each stratum depending on the number of incomplete lists. Formally, let $S = s$ indicates the stratum where $Y_1, \ldots, Y_m$, $m < k$, do not operate. Then, for that stratum we have the following 2^{m+1} structural zero cells:

$$\left\{ n_{x,y_1,\ldots,y_m,0,\ldots,0,s}^{X,Y_1,\ldots,Y_m,Y_{m+1},\ldots,Y_k,S} \right\}_{(x,y_1,\ldots,y_m) \in \{0,1\}^{m+1}},$$

which are to be excluded in the EM algorithm, and are estimated after its convergence.

As an example, consider lists A, B, C, and D where list A is incomplete and $S \in \{s_1, s_2\}$ indicates if A operates or not. The observed marginal counts are $T_1 = [n_{abcds_1}]$ and $T_2 = [n_{bcds_2}]$. The structural zero cells for T^* are:

$$n_{x0000s_1}^{XABCDS}, \qquad n_{x0000s_2}^{XABCDS}, \quad \text{and} \quad n_{x1000s_2}^{XABCDS} \quad \text{for } x \in \{0, 1\}. \tag{8.7}$$

The log-likelihood of the observed incomplete data is:

$$\sum_{a,b,c,d} n_{abcds_1} \log \pi_{abcds_1} + \sum_{b,c,d} n_{bcds_2} \log \pi_{bcds_2},$$

while the EM algorithm is specified in the following way:

---

1. initialize at random an estimate of the conditional probabilities $\{\hat{\pi}_{x|abcds_1}\}$ and $\{\hat{\pi}_{xa|bcds_2}\}$;

2. estimate the complete contingency table $\hat{T}^* = [n_{xabcd}]$ excluding the structural zero cells (8.7) by computing

$$\hat{n}_{xabcds_1} = n_{abcds_1} \hat{\pi}_{x|abcds_1} \qquad \forall \text{ cells s.t. } (a, b, c, d) \neq (0, 0, 0, 0)$$
$$\hat{n}_{xabcds_2} = n_{bcds_2} \hat{\pi}_{xa|bcds_2} \qquad \forall \text{ cells s.t. } (b, c, d) \neq (0, 0, 0);$$

3. compute the MLE for the loglinear model of choice on the current $\hat{T}^*$ conditionally on the unobservable structural zero cells (8.7);

4. update the current estimates of the conditional probabilities:

$$\hat{\pi}_{x|s_1 abcd} = \frac{\hat{n}_{xabcds_1}}{\sum\limits_{a,b,c,d} \hat{n}_{xabcds_1}}, \qquad \hat{\pi}_{xa|s_2bcd} = \frac{\hat{n}_{xabcds_2}}{\sum\limits_{b,c,d} \hat{n}_{xabcds_2}};$$

5. repeat 2-4 until convergence.

After convergence, we estimate cells (8.7).

Note that, in some of the models we are considering, it is possible to employ a simpler approach for the incomplete lists (or missing data in general). In fact, if the model is decomposable, and the missing and observed parts form distinct factors of the likelihood, we can employ what is called a "full information maximum likelihood" approach. In that case, a missing value for a unit simply implies the absence of the relative parameter in the likelihood. For example, if B is an incomplete list (or in any case a list with missing values) in model

$$\pi_{abcds} = \sum_{x \in \{0,1\}} \pi_x \pi_{a|x} \pi_{b|x} \pi_{cd|x} \pi_s.$$

then, all units having a missing value for B, contribute to the likelihood by the quantity

$$\sum_{x} \pi_x \pi_{a|x} \pi_{cd|x},$$

without the necessity of further computational effort. Under the local independence model (8.2), any pattern of missing values can be approached in this way.

Unfortunately, not all missing patterns can be treated as described in this paragraph. In fact, we need a sub-population where all lists operate simultaneously in order to be able to estimate a latent class model. The reason relies on the fact that, even in the simplest case of model (8.3), the minimal sufficient statistics are the observed counts n_y and we can achieve no data reduction via sufficiency (see [12]). Hence, if in our data we cannot observe n_y, but just some marginal counts, our estimates could be biased. Chances are that, in cases where just the higher order joint distributions are not observed, (i.e., in loglinear terms, the higher order interactions cannot be estimated) the influence of those terms is low, and the eventual bias in estimating N_1 is negligible.

8.4   Evaluating the interpretation of the latent classes

The basic idea behind our model is essentially the same lying behind probabilistic Record Linkage problems, where the binary latent variable is meant to identify true and false matches. That is, in our model we are implicitly assuming that there are two sub-populations with different distributions representing the desired cases and the spurious cases, and that a mixture of two distributions would represent our data.

Of course, this interpretation of the latent variable is amenable to validation. In fact, the common use of finite mixture models in capture–recapture is that of modeling unobserved heterogeneity of capture probabilities among individuals which remains unexplained after the inclusion of the available covariates. The rationale is that of enhancing the fitting by modeling distinct sub-populations which are not directly identified by the observed data. Typically, the choice of the number of components is a matter of balance between goodness-of-fit and parsimony in the number of parameters. So, a mixture of two latent components need not necessarily represent our desired subpopulations of interest (in-scope and out-of-scope units). For example, if we have two clearly distinct groups of units with very different capture probabilities with a similar rate of false captures, a two-class model would likely identify these two groups rather than the false and true captures.

Of course, the plausibility of our interpretation of the latent variable can be tested a posteriori. In fact, unless the model is heavily misspecified, the estimated probabilities $\hat{\pi}_{\underline{y}|x}$ and the estimated posterior probabilities $\hat{\pi}_{x|\underline{y}}$ would help us in interpreting the latent variable, in identifying the latent class nature, and understanding if the hypothesis we made was tenable.

Our prior knowledge on the matter should direct us on evaluating the reasonableness of those probabilities. For example, in many cases of interest, we would expect a low false capture rate in almost all sources, and positive interactions between them. If the manifest variables are all highly correlated with X, and the latent classes are well separated, i.e., if the probabilities $\hat{\pi}_{x|\underline{y}}$ are close to zero or one for a class x, then this constitutes evidence in favour of our thesis. In particular, we would expect the probabilities $\hat{\pi}_{1|\underline{1}}$ and $\hat{\pi}_{0|\underline{0}}$ to be close to one. If, on the converse, all sources have a capture probability above 0.5 in both latent classes, and $\hat{\pi}_{1|\underline{1}}$ and $\hat{\pi}_{0|\underline{0}}$ are far from one, this would suggest an interpretation of the latent classes as two sub-populations with different propensity of being captured.

Note that, all finite mixture models are at best identifiable up to an arbitrary permutation of the label of the latent classes (label switching problem). Hence, it is always necessary to identify the latent class representing the true captures by analyzing the estimated parameters, even if the tenability of the model is not an issue.

We note in passing that the correlation of each manifest variable with X can be expressed in terms of under- and overcoverage rates in a simple form:

$$Cov[Y,X] = \pi_{11}^{YX} - \pi_1^X \pi_1^Y$$
$$= \pi_1^X \pi_0^X \left(1 - \pi_{0|1}^{Y|X} - \pi_{1|0}^{Y|X}\right)$$

hence, if the sum of the under- and overcoverage rates of a source Y is close to 1, Y will be, in general, weakly correlated with X, and its contribution to the model would be poor. If the sum of the error rates exceeds 1, the correlation would be negative, and the role of Y would likely be confounding in the identification and interpretation of the latent variable. Thus, if we know that a source is of poor quality, with high error rates, excluding it from the model may be the best choice in some cases.

It is possible to cover both aspects (unobserved heterogeneity in capture probabilities and true/false captures) in the same model by using multiple latent variables or multiple latent classes. The use of multiple latent variables in latent class models was hinted as early as in [14], (see also [16], [4]). Including two (or more) latent variables allows us to specify more complex dependence structures in our models and to tune the interactions in a finer way. However, one should identify the latent variables, their respective number of modalities, and their dependencies in graph G. So, unless we have a specific knowledge of the subject where the additional latent variables reflect a specific hypothesis on the mechanism generating data, a simpler solution is to consider a single latent variable with a number of latent classes greater than two. In that way, one has to select the best number of classes, and then divide the latent classes into two groups by labeling them as representing either the true captures or false captures. Let g be the chosen number of latent classes $x_1, \ldots, x_g$, and let w_1 and w_0 be the sets of classes representing true and false cases, respectively. Then, the posterior probability of belonging to our target population of a unit with a profile $\underline{y}$ would be:

$$\frac{\displaystyle\sum_{i \in w_1} \pi_{x_i, \underline{y}}}{\displaystyle\sum_{i=1}^{g} \pi_{x_i, \underline{y}}}$$

For an application of the two approaches, see [10] for an example in multiple diagnostic tests in medicine, and [18] for one in record linkage.

8.5 A Bayesian approach

In this section we present a Bayesian approach to our models. We will focus on decomposable models only, since a Bayesian approach to these models is straightforward, while the general case presents some computational difficulties.

An obvious advantage of a Bayesian approach is the possibility to include prior knowledge on the data at hand in a simple way. We can set an informative prior to model the belief on the value of a parameter, and, instead of fixing it as seen in Section 8.2.4, we can tune our degree of confidence in a smooth way. In addition, informative priors can help us identify the desired latent sub-populations.

In a Bayesian approach, we can easily compute the interval estimation of the population size, which has always been difficult in capture–recapture studies. In non-Bayesian contexts we are bounded to a bootstrap approach, which is computationally intensive since latent variables are involved. Stanghellini and van der Heijden [31], extending a result of [7], propose a method based on the use of the profile log-likelihood in loglinear models to estimate a confidence interval for the undercount conditionally on the value of the covariates. However, their method cannot be easily extended to the computation of a confidence interval for the total undercount. On the converse, in a Bayesian approach, interval estimates of the population size are naturally obtained as we inspect its posterior distribution.

Another point in favour is the availability of several tools to account for model uncertainty. In particular, we have some results in model averaging and model selection specifically designed for decomposable loglinear models (see [20], [21]).

We have two proposals for a prior distribution of a decomposable model. The first one is a class of priors known as Hyper–Dirichlet which is conjugate to model (8.4) and has the property of being closed under marginalization (see [9]). For each maximal clique $\mathcal{C}$ of, say, h variables, and each distribution $\pi^{\mathcal{C}}$, we set a Dirichlet over the $(h-1)$–dimensional simplex having parameters $\alpha_{\underline{y}_\mathcal{C}}$ defined for each possible combination of values $\underline{y}_\mathcal{C} \in \{0,1\}^h$ of the variables in $\mathcal{C}$. These priors are not independent, so, to ensure that the marginal distribution of any intersection between cliques is consistent, we impose the following restrictions on the parameters:

$$\sum \alpha_{\underline{y}_{\mathcal{C}_{i-1}}} = \alpha_{\underline{y}_{\mathcal{S}_i}} = \sum \alpha_{\underline{y}_{\mathcal{C}_i}}, \qquad \text{for all } i = 2, \ldots, g,$$

where the sums are to be intended over the combinations of variables consistent with $\underline{y}_{\mathcal{S}_i}$. An application of this class of priors can be found in [21].

In the second approach, we refer to expression (8.5). For each $\pi^{\mathcal{C}_i|\mathcal{S}_i}$ and for each fixed value of $\mathcal{S}_i$, we set a Dirichlet distribution with no further restriction. These Dirichlet distributions are independent of each other, as the $\pi^{\mathcal{C}_i|\mathcal{S}_i}$ are independent by construction, and it is not hard to see that this class of priors is conjugate to (8.5). This approach is utilized for example in the Bayesian analysis of latent class models under local independence (8.2) (see, e.g., [41]).

As an illustration, consider model $[ABX][CDX]$ where $\mathcal{C}_1 = \{A, B, X\}$, $\mathcal{C}_2 = \{C, D, X\}$, $\mathcal{S}_2 = \{X\}$. The first approach would set two Dirichlet distributions:

$$\pi^{ABX} \sim Dir(\alpha_{abx}^{ABX}) \quad \text{and} \quad \pi^{CDX} \sim Dir(\alpha_{cdx}^{CDX}),$$

with the restriction:

$$\alpha_x = \sum_{a,b} \alpha_{abx} = \sum_{c,d} \alpha_{cdx},$$

to ensure that π^X would coincide in the two cases. The posteriors

$$\pi^{ABX} \sim Dir(\alpha_{abx} + n_{abx}) \quad \text{and} \quad \pi^{CDX} \sim Dir(\alpha_{cdx} + n_{cdx})$$

would still be consistent since $\sum_{a,b} n_{abx} = \sum_{c,d} n_{cdx}$.

In the second approach we would set four Dirichlet distributions over the conditional distributions:

$$\pi^{AB|X} \sim Dir(\alpha_{ab|x}^{AB|X}) \, (x = 0, 1), \qquad \pi^{CD|X} \sim Dir(\alpha_{cd|x}^{CD|X}) \, (x = 0, 1),$$

and a Beta distribution $\pi^X \sim Beta(\alpha_0^X, \alpha_1^X)$.

Non-informative priors for the Dirichlet are obtained setting all parameters α equal to 1/2 (following Jeffreys prior for multinomial sampling), or to 1 (following a uniform distribution). If, on the converse, we want to include some prior knowledge, we can refer to the usual interpretation of the Dirichlet parameters as "pseudo-counts" to be added to the actual counts n_y. So, for example, if we believe a source Y has little overcoverage, we would set $\alpha_{0|0}^{Y|X}$ to be larger than $\alpha_{1|0}^{Y|X}$.

It remains to set a prior distribution over n_0, or, equivalently, over N. The following are the common options presented in the literature on Bayesian capture–recapture:

- an improper flat prior: $P(N) \propto 1/N$;

- a Poisson distribution, eventually together with a hyper-prior over its parameter: $N \sim Poi(\lambda)$, $\quad \lambda \sim Gamma(\alpha, \beta)$;

- Rissanen's distribution ([29]) which is always proper and is given by $P(N) \propto 2^{-\log^*(N)}$, where $\log^*(N)$ is the sum of the positive terms in the sequence $\{\log_2(N), \log_2(\log_2(N)), \ldots\}$.

We further hypothesize that the prior distributions over N and Θ are independent of each other.

8.5.1 MCMC algorithm

In this section we detail the steps of a Gibbs-based MCMC algorithm to sample from the posterior distribution of N_1. Let us denote as Θ the parameters of the model, whether they are $\{\pi^{\mathcal{C}_i|\mathcal{S}_i}\}$ or $\{\pi^{\mathcal{C}_i}\}$. Then, at iteration $(t+1)$

1. sample all parameter $\Theta^{(t+1)}$ from their posterior conditional distributions which are Dirichlet distributions;

2. sample from $P(N \mid \Theta^{(t+1)}, T)$. Note that

$$P(N \mid \Theta, T) = P(N \mid \Theta, n_{obs}) = \frac{P(N)}{P(n_{obs} \mid \Theta)} P(n_{obs} \mid N, \Theta)$$

$$\propto P(N) \binom{N}{n_{obs}} \pi_{\underline{0}}^{N - n_{obs}} (1 - \pi_{\underline{0}})^{n_{obs}}. \qquad (8.8)$$

Then, if we choose the improper prior $P(N) \propto 1/N$, we have

$$N^{(t+1)} \sim NegBin\left(n_{obs}, 1 - \pi_{\underline{0}}^{(t+1)}\right)$$

3. sample $n_{x,\underline{y}}^{(t+1)}$ from $P(N_{x,\underline{y}} \mid \Theta^{(t+1)}, N^{(t+1)}, T)$ for all $\underline{y}$:

$$N_{x,\underline{y}}^{(t+1)} \sim Bin\left(n_{\underline{y}}, \pi_{x|\underline{y}}^{(t+1)}\right) \quad \text{where} \quad n_{\underline{0}} = N^{(t+1)} - n_{obs}.$$

If we refer to the example of model $[ABX][CDX]$, and to the parameterization (8.5), the first step consists of the following three steps:

- sample $\pi_x^{(t+1)}$ from the posterior $P\left(\pi^X \mid N_{x,\underline{y}}\right)$, which is the Beta distribution $Beta\left(n_x^{(t)} + \alpha_x\right)$ where $n_x^{(t)} = \sum_{\underline{y}} n_{x,\underline{y}}$;

- sample $\pi_{ab|x}^{(t+1)}$ from $P\left(\pi^{AB|X} \mid N_{x,\underline{y}}\right)$, which is the Dirichlet distribution $Dir\left(n_{abx}^{(t)} + \alpha_{ab|x}\right)$ where $n_{abx}^{(t)} = \sum_{\underline{y}:(A=a,B=b)} n_{x,\underline{y}}$;

- sample $\pi_{cd|x}^{(t+1)}$ from $P\left(\pi^{CD|X} \mid N_{x,\underline{y}}\right)$, which is the Dirichlet distribution $Dir\left(n_{cdx}^{(t)} + \alpha_{cd|x}\right)$ where $n_{cdx}^{(t)} = \sum_{\underline{y}:(C=c,D=d)} n_{x,\underline{y}}$;

and, in the third step, we have

$$\pi_{x|\underline{y}} = \frac{\pi_x^{(t+1)} \pi_{ab|x}^{(t+1)} \pi_{cd|x}^{(t+1)}}{\sum_x \pi_x^{(t+1)} \pi_{ab|x}^{(t+1)} \pi_{cd|x}^{(t+1)}}.$$

The above algorithm becomes slightly more involved if a Poisson or Rissanen prior instead of $1/N$ is adopted for N. In particular, we include in step 2 above a Metropolis-Hastings step (Metropolis-Hastings-within-Gibbs algorithm), to sample a value $N^{(t+1)}$ from $\pi(N \mid \Theta, n_{obs})$.

8.5.2   Simulations results

In this section we report the results of a simulation to empirically assess the estimation algorithm.

We considered two scenarios: in the first one (reported in Figure 8.2), the model used to generate the data, and the estimating model have the same interaction cliques $[AX], [BX], [CDX]$. In the second scenario (reported in Figure 8.3), we test the robustness of the model to misspecification by generating the data from a model $[ABX], [CDX]$ and estimating it with the model $[AX], [BX], [CDX]$. For each scenario we generated two datasets with two different population sizes: one with $N = 500$, and one with $N = 1,000,000$. In the generating models of each scenario, we set $\pi_0^X = 0.4$, that is, a proportion of out-of-scope units (both captured and uncaptured) equal to 40%, and a proportion of unobserved units (both in-scope and out-of-scope) equal to 23%. The only difference in Scenario 2 is the presence of an interaction between A and B. In particular, A and B have a correlation of about 0.6 both under $X = 1$ and $X = 0$, while they are conditionally independent in the estimating model.

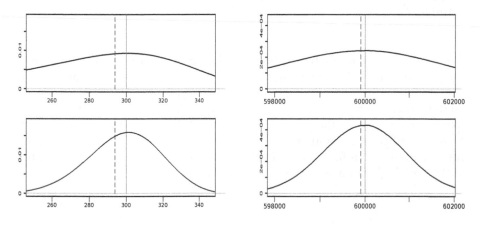

FIGURE 8.2
Posterior distributions of N_1 in Scenario 1. Results for N=500 (left graphs) and N=1,000,000 (right graphs), non-informative priors (top graphs), and informative priors (bottom graphs). The solid line indicates the true value of N_1, the dashed line the Maximum Likelihood Estimate $\hat{N}_1$.

To evaluate the influence of the prior distributions on the estimates, we set two cases. In the first one, we set non-informative priors for all parameters (all Dirichlet parameters equal to 1 and $P(N) = 1/N$). In the second, we mimic an informative scenario coming from an audit sample: we took a 5% sample of the generated complete population [XABCD], and set the parameters of the Dirichlet equal to the observed counts in that sample. In addition, to simulate an informative prior for N, we set a Poisson distribution having

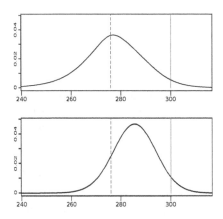
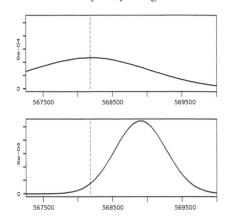

FIGURE 8.3
Posterior distributions of N_1 in Scenario 2. Results for $N=500$ (left graphs) and $N=1,000,000$ (right graphs), non-informative priors (top graphs), and informative priors (bottom graphs). The solid line indicates the true value of N_1 (not visible for $N=1,000,000$), the dashed line the Maximum Likelihood Estimate $\hat{N}_1$.

parameter λ equal to the true value of N. In Figure 8.3 we can see that, when the population size is not too large ($N = 500$), correct informative priors sensibly mitigate the error coming from the missing parameter in the estimating model. When the population size is large ($N=1,000,000$), even if the informative priors influenced the posterior in the right direction, their contribution seems insufficient to compensate for the missing parameter (at least in the case we present). In fact, the real value of N_1 (600,000) is not even included in the 99% highest posterior density interval.

Bibliography

[1] A. Agresti. Simple capture–recapture models permitting unequal catchability and variable sampling effort. *Biometrics*, 50(2):494–500, 1994.

[2] E.S. Allman, J.A. Rhodes, E. Stanghellini, and M. Valtorta. Parameter identifiability of discrete Bayesian networks with hidden variables. *Journal of Causal Inference*, 3(2):189–205, 2015.

[3] S. Asmussen and D. Edwards. Collapsibility and response variables in contingency tables. *Biometrika*, 70(3):567–578, 1983.

[4] P.P. Biemer, R.M. Groves, L.E. Lyberg, N.A. Mathiowetz, and S. Sudman. *Measurement errors in surveys*, volume 173. John Wiley & Sons, 2011.

[5] A. Biggeri, E. Stanghellini, F. Merletti, and M. Marchi. Latent class models for varying catchability and correlation among sources in Capture-Recapture estimation of the size of a human population. *Statistica Applicata*, 11(3):1–14, 1999.

[6] D. Böhning and D. Schön. Nonparametric maximum likelihood estimation of population size based on the counting distribution. *Journal of the Royal Statistical Society: Series C (Applied Statistics)*, 54(4):721–737, 2005.

[7] R.M. Cormack. Interval estimation for mark-recapture studies of closed populations. *Biometrics*, 48(2):567–576, 1992.

[8] J. Daggy, H. Xu, S. Hui, and S. Grannis. Evaluating latent class models with conditional dependence in record linkage. *Statistics in medicine*, 33 (24):4250–4265, 2014.

[9] A.P. Dawid and S.L. Lauritzen. Hyper markov laws in the statistical analysis of decomposable graphical models. *The Annals of Statistics*, 21 (3):1272–1317, 1993.

[10] N. Dendukuri, A. Hadgu, and L. Wang. Modeling conditional dependence between diagnostic tests: a multiple latent variable model. *Statistics in medicine*, 28(3):441–461, 2009.

[11] S.E. Fienberg. The multiple recapture census for closed populations and incomplete 2k contingency tables. *Biometrika*, 59(3):591–603, 1972.

[12] S.E. Fienberg, P. Hersh, A. Rinaldo, and Y. Zhou. Maximum likelihood estimation in latent class models for contingency table data. In P. Gibilisco, editor, *Algebraic and geometric methods in statistics*, pages 27–62. Cambridge University Press, 2010.

[13] H. Glickman, R. Nirel, and D. Ben-Hur. False captures in capture-recapture experiments with application to census adjustment. *Bulletin of the International Statistical Institute, 54th session*, LX:413–414, 2003.

[14] L.A. Goodman. Exploratory latent structure analysis using both identifiable and unidentifiable models. *Biometrika*, 61(2):215–231, 1974.

[15] J.A. Hagenaars. Latent structure models with direct effects between indicators: local dependence models. *Sociological Methods & Research*, 16(3):379–405, 1988.

[16] J.A. Hagenaars. *Loglinear models with latent variables*, volume 94. Sage, 1993.

[17] C.S. Kamen. The 2008 Israel integrated census of population and housing. *Statistical Journal of the United Nations Economic Commission for Europe*, 22(1):39–57, 2005.

[18] M.D. Larsen and D.B. Rubin. Iterative automated record linkage using mixture models. *Journal of the American Statistical Association*, 96(453): 32–41, 2001.

[19] R.J.A. Little and D.B. Rubin. *Statistical analysis with missing data*. John Wiley & Sons, 2014.

[20] D. Madigan and A.E. Raftery. Model selection and accounting for model uncertainty in graphical models using occam's window. *Journal of the American Statistical Association*, 89(428):1535–1546, 1994.

[21] D. Madigan and J.C. York. Bayesian methods for estimation of the size of a closed population. *Biometrika*, 84(1):19–31, 1997.

[22] L. Mancini and S. Toti. Dalla popolazione residente a quella abitualmente dimorante: modelli di previsione a confronto sui dati del censimento 2011, Istat Working Paper vol.8, 2014, ISTAT.

[23] D. Manrique-Vallier. Bayesian population size estimation using Dirichlet process mixtures. *Biometrics*, 72(4):1246–1254, 2016.

[24] A.B. Mooijaart and P.G.M. Van der Heijden. The EM algorithm for latent class analysis with equality constraints. *Psychometrika*, 57(2):261–269, 1992.

[25] D.L. Oberski, G.H. van Kollenburg, and J.K. Vermunt. A Monte Carlo evaluation of three methods to detect local dependence in binary data latent class models. *Advances in Data Analysis and Classification*, 7(3): 267–279, 2013.

[26] D.L. Oberski, J.K. Vermunt, and G.B.D. Moors. Evaluating measurement invariance in categorical data latent variable models with the EPC-Interest. *Political analysis*, 23(4):550–563, 2015.

[27] R. Pradel. Multievent: an extension of multistate capture–recapture models to uncertain states. *Biometrics*, 61(2):442–447, 2005.

[28] Y. Qu, M. Tan, and M.H. Kutner. Random effects models in latent class analysis for evaluating accuracy of diagnostic tests. *Biometrics*, 52(3): 797–810, 1996.

[29] J. Rissanen. A universal prior for integers and estimation by minimum description length. *The Annals of statistics*, 11(2):416–431, 1983.

[30] J.L. Schafer. *Analysis of incomplete multivariate data*. CRC Press, 1997.

[31] E. Stanghellini and P.G.M. van der Heijden. A multiple-record systems estimation method that takes observed and unobserved heterogeneity into account. *Biometrics*, 60(2):510–516, 2004.

[32] E. Stanghellini and B. Vantaggi. Identification of discrete concentration graph models with one hidden binary variable. *Bernoulli*, 19(5A):1920–1937, 2013.

[33] J.M. Sutherland. *Multi-list methods in closed populations with stratified or incomplete information.* PhD thesis, (Dept. of Statistics and Actuarial Science)/Simon Fraser University, 2003.

[34] J.M. Sutherland and C.J. Schwarz. Multi-list methods using incomplete lists in closed populations. *Biometrics*, 61(1):134–140, 2005.

[35] J. Thandrayen and Y. Wang. A latent variable regression model for capture–recapture data. *Computational Statistics & Data Analysis*, 53 (7):2740–2746, 2009.

[36] J. Thandrayen and Y. Wang. Capture–recapture analysis with a latent class model allowing for local dependence and observed heterogeneity. *Biometrical Journal*, 52(4):552–561, 2010.

[37] K. Tilling and J.A.C. Sterne. Capture-recapture models including co-variate effects. *American journal of epidemiology*, 149(4):392–400, 1999.

[38] M. van Smeden, C.A. Naaktgeboren, J.B. Reitsma, K.G.M. Moons, and J.A.H. de Groot. Latent class models in diagnostic studies when there is no reference standard? a systematic review. *American journal of epidemiology*, 179(4):423–431, 2013.

[39] J.K. Vermunt. Log-linear event history analysis: A general approach with missing data, unobserved heterogenity, and latent variables. *Dissertation abstracts international*, 58(1, spring), 1997.

[40] J.K. Vermunt and J. Magidson. Technical guide for latent gold 4.0: Basic and advanced. *Belmont Massachusetts: Statistical Innovations Inc*, 2005.

[41] A. White and T.B. Murphy. Bayeslca: An R package for Bayesian latent class analysis. *Journal of Statistical Software*, 61(13):1–28, 2014.

[42] W.E. Winkler. Methods for adjusting for lack of independence in an application of the fellegi-sunter model of record linkage. *Survey Methodology*, 15(1):101–117, 1989.

[43] E.M. Zwane and P.G.M. van der Heijden. Population estimation using the multiple system estimator in the presence of continuous covariates. *Statistical Modelling*, 5(1):39–52, 2005.

[44] E.N. Zwane, K. van der Pal-de Bruin, and P.G.M. van der Heijden. The multiple-record systems estimator when registrations refer to different but overlapping populations. *Statistics in Medicine*, 23(14):2267–2281, 2004.

9

Log-linear models of erroneous list data

Li-Chun Zhang

University of Southampton, Statistics Norway & University of Oslo

CONTENTS

9.1 Introduction

Log-linear models are widely applied in biological, social and medical studies of capture-recapture data; see, e.g., Fienberg (1972), Wolter (1986), Cormack (1989), IWGDMF (1995a, 1995b), Böhning et al. (2017). A key assumption is that none of the records erroneously enumerates the target population units. However, in many situations, erroneous enumeration, or spurious out-of-scope records, are unavoidable.

Example 1 *Suppose there are several different screening tests for a given disease. Suppose those persons with one or several test results exceeding some prescribed threshold values are subjected to a comprehensive clinical examination, after which a diagnosis of the disease may or may not be the case. Each test can have false positive and false negative results, where the former represents erroneous enumeration of the target patient population. The final clinical examination* *can have false negatives. However, since every diagnosed patient will be treated, one may disregard the false positives in this regard. An analysis of the effectiveness of the initial tests and the final diagnosis should ideally take into account the various over- and under-enumerations.*

Example 2 *Suppose several administrative registers are available for the usual resident population. However, each register suffers from over- and under-enumeration, so that neither the union nor the joint subset of them can provide the population count directly. Suppose a population coverage survey is carried out in addition, which suffers from under-enumeration. This is a relevant setting in many countries, which currently seek alternative approaches to the traditional population census for the provision of population counts.*

Formally, we consider log-linear modelling of *list-survey* capture data under the following set-up. Let the target population $U = \{1, 2, ..., N\}$ have unknown size N. Let there be K enumeration *lists* of U, denoted by A_k for $k = 1, ..., K$, where $A_k \setminus U \neq \emptyset$ and $U \setminus A_k \neq \emptyset$. In addition, let S be an enumeration *survey* of U, where $S \subset U$. The set-up is asymmetric between the lists and the survey, in the sense that erroneous enumeration is present in the enumeration lists, where $A_k \setminus U \neq \emptyset$, but not in the enumeration survey. A 'symmetric' set-up, where erroneous enumeration is present in *all* the lists and survey, is considered in Chapter 8.

Previously, Zhang and Dunne (2017) developed the trimmed dual system estimation in the case of $K = 1$. Zhang (2015) examined various modelling options of erroneous list data in the case of $K = 2$. Two classes of models can be formulated generally. The first one is derived from the standard log-linear models of the cross-classified cells of a contingency table, which is based on the conditional independence type of assumptions. The second class are log-linear models of error probabilities in the *marginally classified* list domains. In the case of $K = 2$, a possible unsaturated model is given by

$$\Pr(i \notin U | i \in A_1, i \in A_2) = \Pr(i \notin U | i \in A_1)\Pr(i \notin U | i \in A_2) \qquad (9.1)$$

i.e., the *joint conditional* probability is the product of the *marginal conditional* probabilities. Zhang (2015) refers to (9.1) as an assumption of *pseudo-conditional independence (PCI)*. It is demonstrated that the two classes of models are complementary to each other, thereby extending the range of data that can be accommodated by the log-linear modelling approach.

Our aim here is to develop and elaborate the PCI type of models for the general case where $K \geq 2$. First, the basic notations are introduced in relation to the standard log-linear models of contingency tables. Next, we define the PCI type of models for the general case, where the model fitting procedures are explained given census (i.e., U) or survey (i.e., $S \subset U$) data. Finally, model selection in cases of zero degree of freedom in the data is discussed, and the approach is illustrated using both simulations and real-life data.

9.2   Log-linear models of incomplete contingency tables

Let $\delta_k(i) = 1$ if unit i is in list k and 0 if it is not. Equivalently, let

$$\omega = \{k; \delta_k(i) = 1, 1 \le k \le K\}$$

be the index set of the lists in which the unit i is enumerated. Let A_ω be the cross-classified *list domain* according to $\delta_1, ..., \delta_K$. Let Ω_K be the set of ω, which consists of all the non-empty subsets of $\{1, ..., K\}$. Together $\{A_\omega; \omega \in \Omega_K\}$ form a cross-classified partition of the *list universe* $A = \bigcup_{k=1}^{K} A_k$, which is of the size $x_A = |A| = \sum_{k=1}^{K} x_\omega$, where $x_\omega = |A_\omega|$ is the observed size of A_ω. Let $\delta_U(i) = 1$ if $i \in U$ and 0 otherwise. Let $y_\omega = \sum_{i \in A_\omega} \delta_U(i)$ be the number of population units in A_ω, which we refer to as the list *hits*; let $r_\omega = x_\omega - y_\omega$ be the list *errors* in A_ω. It is important to note the difference to the notation A_k in (9.1), which is used to designate the k-th list which contains *all* the units in it, whereas A_ω with $\omega = \{1\}$ refers to the list domain which consists of the units *only* enumerated in the k-th list. Later on, we will extend the notation, whereby A_{1+} will be used to designate the set of all the units in the first list.

Consider the contingency table arising from cross-classifying the units in $A \cup U$. Denote by $\mu_{\omega \delta_U} = \mu_{\delta_1 \cdots \delta_K \delta_U}$ the expected count of cross-classified units in $A \cup U$, where ω replaces $\delta_1, ..., \delta_K$, consisting of those k's with $\delta_k = 1$. The most general log-linear model of $\{\mu\}$ can now be given as

$$\log \mu_{\omega \delta_U} = \lambda + \sum_{v \in \Omega(\omega)} \lambda_{1_v}^{A_v} + \lambda_1^U + \sum_{v \in \Omega(\omega)} \lambda_{1_v 1}^{A_v U}$$

where $\Omega(\omega)$ consists of all the non-empty subsets of ω, and as the parameter constraints we set $\lambda_{\omega \delta_U}$ to 0 if there is at least one 0 among $\delta_1 \cdots \delta_K \delta_U$. For example, if $\omega = \{1, 2, 3\}$, then $\lambda_{1010}^{A_1 A_2 A_3 U} = 0$, whereas $\delta_{11}^{A_1 A_3} \ne 0$, which corresponds to $\lambda_{1_v}^{A_v}$ with $v = \{1, 3\}$ as a part of the second term on the right-hand side, or $\delta_{111}^{A_1 A_3 U} \ne 0$, which corresponds to $\lambda_{1_v 1}^{A_v U}$ with $v = \{1, 3\}$ as a part of the fourth term. In this way, we have

$$\log E(y_\omega) = \log \mu_{\omega 1} = \lambda + \sum_{v \in \Omega(\omega)} \lambda_{1_v}^{A_v} + \lambda_1^U + \sum_{v \in \Omega(\omega)} \lambda_{1_v 1}^{A_v U}$$

$$\log E(r_\omega) = \log \mu_{\omega 0} = \lambda + \sum_{v \in \Omega(\omega)} \lambda_{1_v}^{A_v}$$

A logistic model of the hit probability $\xi_\omega = \Pr(i \in U | i \in A_\omega)$ follows as

$$\text{logit } \xi_\omega = \log \mu_{\omega 1} - \log \mu_{\omega 0} = \sum_{v \in \Omega(\omega)} \lambda_{1_v 1}^{A_v U} \tag{9.2}$$

where $\omega \in \Omega_K$. A model of the list error probability $\theta_\omega = 1 - \xi_\omega$ is equally given by (9.2), simply with logit $\theta_\omega = -$logit ξ_ω. In particular, the redundant term λ_1^U has been removed in (9.2), since it is meaningless to speak of list hits (or errors) among units that are not in A. Moreover, retaining λ_1^U (i.e., instead of another term) would impose an undue constraint on the size of $U \setminus A$, which Zhang (2015) refers to as an incident model assumption. This leaves $2^K - 1$ parameters in the saturated model (9.2), i.e., $\{\lambda_{1_\omega^U 1}^{A_\omega^U}; \omega \in \Omega_K\}$.

For various unsaturated models of (9.2), we restrict our attention to the hierarchical models. For instance, given $K = 2$, the largest unsaturated model is logit $\xi_{12} =$ logit $\xi_1 +$ logit ξ_2, where ξ_1 is the hit (or error) probability among the units in list one but not two, and ξ_2 that among the units in list two but not one. As explained in Zhang (2015), there are limitations to how well this model can fit in situations where the quality of the enumeration lists are high. For example, let $\mu_{121} = x_{12} = 10$, $x_1 = x_2 = 1$ and $\mu_{11} = \mu_{21} = 0$, which yields the error probabilities $\xi_1 = \xi_2 = 1$ and $\xi_{12} = 0$. While this cannot be accommodated by logit $\xi_{11} =$ logit $\xi_1 +$ logit ξ_2, it would nearly satisfy (9.1), now that $\Pr(i \notin U | i \in A_{12}) = 0$, $\Pr(i \notin U | i \in A_1 \cup A_{12}) = 1/(x_{12} + x_1)$ and $\Pr(i \notin U | i \in A_2 \cup A_{12}) = 1/(x_{12} + x_2)$. Introducing the PCI type of models can therefore extend the range of applications of the modelling approach.

9.3   Modelling marginally classified list errors

9.3.1   The models

To express the various PCI assumptions, we need a notation for the marginally classified domains of a contingency table, instead of the cross-classified cells. Let $A_{\omega+} = \sum_{v;\omega \subseteq v} A_v$ be the union of all the cross-classified domains up to the margin specified by ω. Table 9.1 details the notations in the case of $K = 3$. For example, if $\omega = \{1,3\}$, then $A_{\omega+} = A_{13} \cup A_{123}$ which includes all the units that belong to both lists one and three, of which A_{13} are the ones not belonging to list two, and A_{123} the ones belonging to list two in addition; similarly, if $\omega = \{3\}$, then $A_{\omega+} = A_3 \cup A_{13} \cup A_{23} \cup A_{123}$.

Let $\theta_{\omega+} = \Pr(i \notin U | i \in A_{\omega+})$, for $\omega \in \Omega_K$, such that $E(r_{\omega+} | x_{\omega+})$ is the expected errors in the marginal list domain $A_{\omega+}$. The most general log-linear model of $\theta_{\omega+}$ can be given as

$$\log \theta_{\omega+} = \eta_\omega = \sum_{v \in \Omega(\omega)} \alpha_v \qquad (9.3)$$

It is well defined since each of the $2^K - 1$ error probabilities $\{\theta_{\omega+}\}$ is allowed its own distinct parameter, i.e., α_ω for $\theta_{\omega+}$, such that the saturated model always fits the data exactly, provided binomially distributed $\{r_{\omega+}\}$.

TABLE 9.1

Cross vs. marginal classification, $K = 3$.

Cross-classification		Marginal Classification	
$(\delta_1, \delta_2, \delta_3)$	ω	$(\delta_1, \delta_2, \delta_3)$	$\omega+$
$(1, 1, 1)$	123	$(1, 1, 1)$	123+
$(1, 1, 0)$	12	$(1, 1, -)$	12+
$(1, 0, 1)$	13	$(1, -, 1)$	13+
$(0, 1, 1)$	23	$(-, 1, 1)$	23+
$(1, 0, 0)$	1	$(1, -, -)$	1+
$(0, 1, 0)$	2	$(-, 1, -)$	2+
$(0, 0, 1)$	3	$(-, -, 1)$	3+
$(0, 0, 0)$	$\emptyset$		

Alternatively, but not equivalently, a model of the hit probabilities can be given as

$$\log \xi_{\omega+} = \sum_{v \in \Omega(\omega)} \beta_v$$

Or, using the logit link function instead, a model can be given as

$$\text{logit } \theta_{\omega+} = \eta_\omega = \sum_{v \in \Omega(\omega)} \alpha_v = -\text{logit } \xi_{\omega+} \tag{9.4}$$

Each PCI type assumption corresponds to an unsaturated model (9.3). For instance, given $K = 2$, the largest unsaturated model (9.3) amounts to

$$\theta_{12+} = e^{\alpha_1} e^{\alpha_2} = \theta_{1+} \theta_{2+}$$

which is precisely the PCI assumption (9.1). The corresponding unsaturated logistic model (9.4) defines the log odds of error against hit. The link functions logit and log are close to each other, when the odds are low. Together they extend the range of application of the modelling approach.

TABLE 9.2

Illustration of log-linear models (9.3) of list error probabilities, $K = 3$.

Model Restrictions	Model Interpretation
-	Saturated model
$\alpha_{123} = 0$	Null 2nd-order PCI-interaction
$\alpha_{12} = \alpha_{123} = 0$	PCI between A_{1+} and A_{2+} given A_{3+}
$\alpha_{12} = \alpha_{13} = \alpha_{123} = 0$	PCI between A_{1+} and (A_{2+}, A_{3+})
$\alpha_{12} = \alpha_{13} = \alpha_{23} = \alpha_{123} = 0$	Mutual PCI between A_{1+}, A_{2+} and A_{3+}

To appreciate the various PCI type assumptions, and the analogy to the class of hierarchical log-linear models of the cross-classified contingency tables, let us consider more closely the case of $K = 3$, as summarised in Table 9.2. The model (9.3) specifies 7 error probabilities, one for each marginal list domain.

- As an analogy to mutual independence, let the *mutual* PCI assumption of errors between A_{1+}, A_{2+} and A_{3+} be given by

$$\Pr(i \notin U | i \in A_{1+}, i \in A_{2+}, i \in A_{3+})$$
$$= \Pr(i \notin U | i \in A_{1+})\Pr(i \notin U | i \in A_{2+})\Pr(i \notin U | i \in A_{3+})$$

i.e., $\theta_{123+} = \theta_{1+}\theta_{2+}\theta_{3+}$, which corresponds to the restrictions

$$\alpha_{12} = \alpha_{13} = \alpha_{23} = \alpha_{123} = 0$$

i.e., the same form of restrictions for mutual independence under log-linear models of cross-classified cells of a contingency table. Moreover, we have

$$\theta_{12+} = \theta_{1+}\theta_{2+} \quad \text{and} \quad \theta_{13+} = \theta_{1+}\theta_{3+} \quad \text{and} \quad \theta_{23+} = \theta_{2+}\theta_{3+}$$

such that mutual PCI induces pairwise PCI, in the same way as mutual independence induces pairwise independence.

- Similarly, the model of PCI between A_{1+} and (A_{2+}, A_{3+}) is given by

$$\alpha_{12} = \alpha_{13} = \alpha_{123} = 0 \quad \Rightarrow \quad \theta_{123+} = \theta_{1+}\theta_{23+}$$

which induces $\theta_{12+} = \theta_{1+}\theta_{2+}$ and $\theta_{13+} = \theta_{1+}\theta_{3+}$ in addition.

- As another generalisation of pairwise PCI, the errors in $A_{\nu+}$ and $A_{\psi+}$ are said to be *PCI given $A_{\omega+}$*, where $\nu \cap \psi = \nu \cap \omega = \psi \cap \omega = \emptyset$, provided

$$\frac{\theta_{\nu \cup \omega \cup \psi+}}{\theta_{\omega+}} = \left(\frac{\theta_{\nu \cup \omega+}}{\theta_{\omega+}}\right)\left(\frac{\theta_{\psi \cup \omega+}}{\theta_{\omega+}}\right) \tag{9.5}$$

Let $\nu = \{1\}$, $\psi = \{2\}$ and $\omega = \{3\}$. Then, provided the restrictions

$$\alpha_{12} = \alpha_{123} = 0$$

we have

$$\log \theta_{123+} - \log \theta_{3+} = (\alpha_1 + \alpha_2 + \alpha_3 + \alpha_{13} + \alpha_{23}) - \alpha_3$$
$$= (\alpha_1 + \alpha_3 + \alpha_{13} - \alpha_3) + (\alpha_2 + \alpha_3 + \alpha_{23} - \alpha_3)$$
$$= \left(\log \theta_{13+} - \log \theta_{3+}\right) + \left(\log \theta_{23+} - \log \theta_{3+}\right)$$

i.e., PCI between A_{1+} and A_{2+} given A_{3+}. Notice that the three terms $\theta_{123+}/\theta_{3+}$, θ_{13+}/θ_{3+} and θ_{23+}/θ_{3+} are *not* conditional probabilities, except that the six probabilities involved here obey the same calculus as the probabilities satisfying conditional independence.

- Finally, the largest unsaturated model (9.3) is given by the null 2nd-order *PCI-interaction* assumption

$$\alpha_{123} = 0$$

in analogy to the hierarchical log-linear model of contingency tables.

9.3.2 Maximum likelihood estimation

Maximum likelihood estimation (MLE) of $\{\theta_{\omega+}\}$ under the model (9.3) or (9.4) is straightforward in conception, provided one has observations of binomial variables $r_\omega \sim \text{binomial}(x_\omega, \theta_\omega)$, in each cross-classified list domain A_ω. (In case of a random sample from A, one only needs to replace $\{x_\omega\}$ and $\{r_\omega\}$ by the corresponding observed domain sizes and errors.) The log-likelihood is then given as

$$\ell(\boldsymbol{\alpha};\mathbf{r}) = \sum_{\omega \in \Omega_K} r_\omega \log \theta_\omega + y_\omega \log(1 - \theta_\omega)$$

where $y_\omega = x_\omega - r_\omega$. The cross-classified error probability vector $\boldsymbol{\theta}$ can be obtained from the marginally specified $\boldsymbol{\theta}_+$ via the linear relationship

$$E(r_{\omega+}) = \sum_{\nu;\omega \subseteq \nu} E(r_\nu)$$

where $\boldsymbol{\theta}_+$ is specified either by (9.3) or (9.4). Let the two vectors $\boldsymbol{\theta}$ and $\boldsymbol{\theta}_+$ be arranged in a corresponding manner, denoted by

$$\vec{\Omega}_2 = \{\{1\}, \{2\}, \{1, 2\}\}$$
$$\vec{\Omega}_3 = \{\{1\}, \{2\}, \{3\}, \{1, 2\}, \{1, 3\}, \{2, 3\}, \{1, 2, 3\}\}$$

for Ω_2 and Ω_3. Arrange $\mathbf{x}$ and $\mathbf{x}_+$ accordingly. We have, then,

$$\boldsymbol{\theta} = \mathbf{C}\boldsymbol{\theta}_+ \qquad \text{for} \quad \mathbf{C} = \text{Diag}(\mathbf{x})^{-1} \, \boldsymbol{\Gamma} \, \text{Diag}(\mathbf{x}_+)$$

where, for $K = 2$,

$$\boldsymbol{\Gamma} = \begin{pmatrix} 1 & 0 & -1 \\ 0 & 1 & -1 \\ 0 & 0 & 1 \end{pmatrix}$$

and, for $K = 3$,

$$\boldsymbol{\Gamma} = \begin{pmatrix} 1 & 0 & 0 & -1 & -1 & 0 & 1 \\ 0 & 1 & 0 & -1 & 0 & -1 & 1 \\ 0 & 0 & 1 & 0 & -1 & -1 & 1 \\ 0 & 0 & 0 & 1 & 0 & 0 & -1 \\ 0 & 0 & 0 & 0 & 1 & 0 & -1 \\ 0 & 0 & 0 & 0 & 0 & 1 & -1 \\ 0 & 0 & 0 & 0 & 0 & 0 & 1 \end{pmatrix}$$

The MLE of $\boldsymbol{\alpha}$ can now be obtained by the Newton-Raphson method. The first- and second-order derivatives are given as follows. The score of α_ω, i.e., the corresponding component of $\boldsymbol{\alpha}$, is given by

$$\frac{\partial \ell}{\partial \alpha_\omega} = \mathbf{d}^\top \left(\frac{\partial \boldsymbol{\theta}}{\partial \alpha_\omega} \right) = \mathbf{d}^\top \left(\mathbf{C} \frac{\partial \boldsymbol{\theta}_+}{\partial \boldsymbol{\eta}} \frac{\partial \boldsymbol{\eta}}{\partial \alpha_\omega} \right) = \mathbf{d}^\top \mathbf{C} \, \text{Diag}(\mathbf{w}) \, \boldsymbol{\Delta}_\omega$$

where Δ_ω is the corresponding column of Δ, for $\Delta^\top = \Gamma^{-1}$ and $\eta = \Delta\alpha$ under both (9.3) and (9.4), and the vector $\mathbf{d}$ of derivatives has components

$$d_\omega = \frac{\partial \ell}{\partial \theta_\omega} = \frac{r_\omega}{\theta_\omega} - \frac{y_\omega}{1 - \theta_\omega}$$

and $\partial \theta_+/\partial \eta = \mathrm{Diag}(\mathbf{w})$, and the components of the vector $\mathbf{w}$ are $w_\omega = \theta_\omega$ under the log model (9.3) and $w_\omega = \theta_\omega(1 - \theta_\omega)$ under the logit model (9.4). The vector of scores is then given by

$$\frac{\partial \ell}{\partial \alpha} = \left(\frac{\partial \theta}{\partial \alpha}\right)^\top \mathbf{d} = \left(\frac{\partial \theta_\Omega}{\partial \alpha}\right)^\top \mathbf{C}^\top \mathbf{d} = \Delta^\top \mathrm{Diag}(\mathbf{w}) \mathbf{C}^\top \mathbf{d}$$

On further derivation, we obtain the matrix of second-order derivatives

$$\frac{\partial^2 \ell}{\partial \alpha \partial \alpha^\top} = \left(\frac{\partial \theta}{\partial \alpha}\right)^\top \mathrm{Diag}(\mathbf{b}) \left(\frac{\partial \theta}{\partial \alpha}\right) + \mathbf{D}$$

where the vector $\mathbf{b}$ of partial derivatives has components

$$b_\omega = \frac{\partial d_\omega}{\partial \theta_\omega} = -\frac{r_\omega}{\theta_\omega^2} + \frac{y_\omega}{(1 - \theta_\omega)^2}$$

and the column of $\mathbf{D}$ corresponding to ω is given by

$$\mathbf{D}_\omega = \Delta^\top \left(\frac{\partial \mathrm{Diag}(\mathbf{w})}{\partial \alpha_\omega}\right) \mathbf{C}^\top \mathbf{d} = \Delta^\top \mathrm{Diag}(\mathbf{v}) \mathrm{Diag}(\Delta_\omega) \mathbf{C}^\top \mathbf{d}$$

and the vector $\mathbf{v}$ of partial derivatives has components $v_\omega = \partial w_\omega/\partial \eta_\omega$, with $v_\omega = \theta_\omega$ under (9.3) and $v_\omega = \theta_\omega(1 - \theta_\omega)(1 - 2\theta_\omega)$ under (9.4).

9.3.3 Estimation based on list-survey data

In many applications (e.g., Example 2) the observed population units may arise from a *separate* coverage survey from the target population, which is subjected to under-enumeration. For $i \in U$, let $\delta_S(i) = 1$ if $i \in S$ and 0 otherwise. Let $n_\omega = |A_\omega \cap S|$ be the number of survey captures in list domain A_ω. The list hits and errors, i.e., (y_ω, r_ω) are not directly observed themselves. Moreover, the out-of-list sub-population $U_\emptyset = U \setminus A$ is partially enumerated, yielding $n_\emptyset$ observed survey captures and $m_\emptyset$ unobserved survey misses, where $N_\emptyset = n_\emptyset + m_\emptyset$. The hypothetical complete data consists of two parts:

- the list errors r_ω (or hits y_ω), for $\omega \in \Omega_K$;

- the survey captures n_ω in A, for $\omega \in \Omega_K$, and $n_\emptyset$ in $U_\emptyset$.

We assume the survey captures follow the IID Bernoulli distribution with constant catch probability $\gamma = \mathrm{Pr}(i \in S | i \in U)$ everywhere. In practice, the assumption can be combined with appropriate stratification of $A \cup U$, provided it is feasible. Given x_ω, the list hits and survey captures constitute two-phase binomial selection from A_ω, with probabilities ξ_ω and γ, respectively. For the out-of-list survey captures $n_\emptyset$, we consider two options below.

- **Binomial model** Let $n_\emptyset$ follow the binomial distribution $(N_\emptyset, \gamma)$, where $N_\emptyset$ is regarded as a fixed unknown parameter.

- **Negative binomial model** Let $m_\emptyset$ follow the negative binomial distribution $(n_\emptyset, 1 - \gamma)$ given $n_\emptyset$, as if $n_\emptyset$ were fixed in advance. In particular, $N_\emptyset$ is then regarded as a random variable. Compared to the binomial model, this entails the loss of one degree of freedom in the data, while equally reducing the number of unknown parameters by one. Finally, the best predictor (BP) of $N_\emptyset$ is given by $E(N_\emptyset|n_\emptyset) = n_\emptyset/\gamma$.

Let $\psi = (\gamma, \alpha)$, with α under (9.3) or (9.4). Under the binomial model, the likelihood based on the observed data $(\mathbf{n}, n_\emptyset)$ is given by

$$L_1(\psi, N_\emptyset) \propto L \cdot L_2$$

$$L(\psi) \propto \prod_{\omega \in \Omega_K} p_\omega^{n_\omega} (1 - p_\omega)^{x_\omega - n_\omega} \tag{9.6}$$

$$L_2(\gamma, N_\emptyset) \propto \binom{N_\emptyset}{n_\emptyset} \gamma^{n_\emptyset} (1 - \gamma)^{N_\emptyset - n_\emptyset}$$

where $p_\omega = \gamma(1 - \theta_\omega) = \Pr(i \in S | i \in A_\omega)$. Whereas, under the negative binomial model, the likelihood based on the observed data $(\mathbf{n}, n_\emptyset)$ is simply $L(\psi)$, on integrating out $m_\emptyset$ from the likelihood based on $(\mathbf{n}, n_\emptyset, m_\emptyset)$, which depends on the data only via $\mathbf{n}$. However, it is well known that, given γ, L_2 is maximised at $\hat{N}_\emptyset = n_\emptyset/\gamma$, without rounding to an integer, which is just the BP of $N_\emptyset$ under the negative binomial model. It follows that maximising L_2 in addition under the binomial model is equivalent to maximising $L(\psi)$ *and* using the BP of $N_\emptyset$ under the negative binomial model. In other words, there is no practical difference which model one assumes.

It is convenient to apply the EM algorithm to maximise (9.6) since we can reuse the MLE procedure described in Section 9.3.2. The complete-data log-likelihood corresponding to $L(\psi)$ is based on $\mathbf{r}$ and $\mathbf{n}$, and is given by

$$\ell_C(\psi; \mathbf{r}, \mathbf{n}) = \sum_{\omega \in \Omega_K} r_\omega \log \theta_\omega + y_\omega \log(1 - \theta_\omega)$$

$$+ n_A \log \gamma + (y_A - n_A) \log(1 - \gamma)$$

where $n_A = \sum_{\omega \in \Omega_K} n_\omega$ and $y_A = \sum_{\omega \in \Omega_K} y_\omega$. The EM algorithm follows.

E-step Given the current parameter values, evaluate the conditional expectation of $\mathbf{y}$ and $\mathbf{r}$ given n_ω, i.e., the unobserved list hits and errors:

$$E(y_\omega|n_\omega) = (x_\omega - n_\omega)(1 - \theta_\omega)(1 - \gamma)/(1 - p_\omega)$$

$$E(r_\omega|n_\omega) = x_\omega - E(y_\omega|n_\omega)$$

M-step Update the estimate of γ by $n_A/E(y_A|\mathbf{n})$, and those of α using the procedure described in Section 9.3.2 based on $E(\mathbf{r}|\mathbf{n})$ and $E(\mathbf{y}|\mathbf{n})$.

On convergence, we obtain the MLE $\hat{\alpha}$ and $\hat{\gamma}$, and the empirical BP of N by

$$\hat{N} = n_A + n_\emptyset + n_\emptyset(1 - \hat{\gamma})/\hat{\gamma} = n_A + n_\emptyset/\hat{\gamma} \tag{9.7}$$

9.4   Model selection with zero degree of freedom

9.4.1   Latent likelihood ratio criterion

Within each class of log or logit models, model selection between two nested models can make use of both the goodness-of-fit and the differing number of parameters, e.g., by the Akaike Information Criterion. Between them there is an issue of choosing one of two models with the same number of parameters but different link functions, in which case the choice formally can only depend on the goodness-of-fit. However, when there are few lists available, it may happen that both the models 'use up' all the degrees of freedom in the data, and both fit the data exactly. For instance, in the case of $K = 2$, the likelihood (9.6) is based on 3 freely varying observations (n_1, n_2, n_{12}). The survey catch probability γ being a necessary parameter, the largest unsaturated list-error model is given by setting $\alpha_{12} = 0$ either in (9.3) or (9.4). This leaves zero degree of freedom for testing the goodness-of-fit formally. Moreover, it can happen that both the models fit the data exactly, so that even informally it is impossible to tell which one is the better-fitting. Below we elaborate a *latent likelihood ratio criterion*, originally discussed in Zhang (1996), which may be relevant for model selection in these circumstances.

The starting point is the standard theory of model selection based on the likelihood ratio test (LRT). See e.g., Vuong (1989) for an account of the general properties of the LRT based on independent and identically distributed (IID) observations. In the present context, the complete data $(\mathbf{r}, \mathbf{n})$ arises from IID realisations of (δ_U, δ_S) for all $i \in A_\omega$ and $\omega \in \Omega_K$, where $\mathbf{r}$ is missing and only $\mathbf{n}$ is observed. The observed data $\mathbf{n}$ is still generated by IID Bernoulli trials with probability p_ω in each A_ω. The likelihood L_1 based on $\mathbf{n}$ is given by (9.6), under either the *working* model (9.3) or (9.4). Let the two models have the same α-terms and only differ in the link function.

Without assuming a working model is necessarily true, the MLE $\hat{\psi}$ under it converges to the so-called *pseudo-true* values, denoted by $\psi^* = (\alpha^*, \gamma^*)$, which are the parameter values of the working model that minimise the *deviance*, i.e., Kullback-Leibler divergence (Vuong, 1989). The pseudo-true values can equivalently be defined as the MLE derived from the *true* expectation of the working-model log-likelihood, which in our case can be given by

$$\psi^* = \arg\min_\psi E^0[\ell(\psi; \mathbf{n})] = \arg\min_\psi \ell(\psi; \mathbf{n}^0)$$

where $\mathbf{n}^0 = E^0(\mathbf{n})$ is the expectation of $\mathbf{n}$ w.r.t. the true distribution of the data, and the last equality holds whenever the log-likelihood is linear in $\mathbf{n}$, which is the case for any model of the exponential family of distributions.

Now, to distinguish between the two models (9.3) and (9.4), both with the same constraints on α, we shall denote them by ψ_{log} and ψ_{logit}, respectively.

The LRT selection is indeterminate between the two models if

$$\ell(\psi^*_{log};\mathbf{n}^0) = \ell(\psi^*_{logit};\mathbf{n}^0) = \ell(\mathbf{n}^0;\mathbf{n}^0) \tag{9.8}$$

where $\ell(\mathbf{n}^0;\mathbf{n}^0)$ denotes the saturated log-likelihood that minimises the deviance by definition. Provided (9.8), Vuong (1989) refers to the two models as equivalent. However, equivalence as such is impractical, as long as the two models give different estimates of N. Moreover, the equivalence is not genuine but due to the missing of $\mathbf{r}$, since the two models could have been distinguished otherwise, as both ψ_{log} and ψ_{logit} are identifiable given $\mathbf{r}$.

One may consider, in the case of (9.8), a secondary robustness criterion and a simulation-based approach to model selection. To start with, given the missing data situation and the exponential family of distributions of the complete data $(\mathbf{r},\mathbf{n})$, the MLE by the EM algorithm is such that

$$E(\mathbf{r}|\mathbf{n};\hat{\psi}_{log}) = \hat{\mathbf{r}} = E(\mathbf{r};\hat{\alpha}_{log})$$

where the first equation corresponds to the calculation at the E-step, and the second one is the score equation that is being solved at the M-step. Provided the consistency of the MLE (Dempster et al., 1977; Wu, 1983), asymptotically as $x_\omega \to \infty$ for each $\omega \in \Omega_K$, we have

$$x^{-1}_\omega E(r_\omega|n_\omega;\hat{\psi}_{log}) \xrightarrow{P} x^{-1}_\omega E(r_\omega|n_\omega;\psi^*_{log}) = x^{-1}_\omega E(r_\omega;\alpha^*)$$

In other words, under model ψ_{log}, the estimated conditional expectation of $\mathbf{r}$ converges to the hypothetical complete data $E(\mathbf{r};\psi^*_{log})$ that would have yielded the pseudo-true ψ^*_{log} as the MLE based on the complete-data log-likelihood. We refer to $E(\mathbf{r};\psi^*_{log})$ as the *pseudo-true* missing data under model ψ_{log}. Similarly, let $E(\mathbf{r};\psi^*_{logit})$ be the pseudo-true missing data under model ψ_{logit}. Meanwhile, the true conditional expectation of r_ω is such that

$$x^{-1}_\omega E^0(r_\omega|n_\omega) \xrightarrow{P} x^{-1}_\omega E^0(r_\omega|n^0_\omega) = x^{-1}_\omega E^0(r_\omega)$$

Generally, where none of the two models is true, the pseudo-true data under either model are not equal to each other, nor the true $E^0(r_\omega)$, i.e.,

$$E(r_\omega|n_\omega;\psi^*_{log}) \neq E(r_\omega|n_\omega;\psi^*_{logit}) \neq E^0(r_\omega|n_\omega)$$

Provided (9.8), we propose a latent likelihood ratio criterion (LRC), whereby model ψ_{log} is deemed more *robust* than model ψ_{logit} provided

$$\ell_C(\mathbf{n},\mathbf{r}^*_{logit};\mathbf{n},\mathbf{r}^*_{logit}) - \ell_C(\hat{\psi}_{log};\mathbf{n},\mathbf{r}^*_{logit})$$
$$\leq \ell_C(\mathbf{n},\mathbf{r}^*_{log};\mathbf{n},\mathbf{r}^*_{log}) - \ell_C(\hat{\psi}_{logit};\mathbf{n},\mathbf{r}^*_{log}) \tag{9.9}$$

where $\mathbf{r}^*_{logit} = E(\mathbf{r}|\mathbf{n}; \boldsymbol{\psi}^*_{logit})$ and $\mathbf{r}^*_{log} = E(\mathbf{r}|\mathbf{n}; \boldsymbol{\psi}^*_{log})$, and $\ell_C(\mathbf{n}, \mathbf{r}^*_{logit}; \mathbf{n}, \mathbf{r}^*_{logit})$ is the saturated complete-data log-likelihood based on the pseudo-true complete data under model $\boldsymbol{\psi}_{logit}$, and $\ell_C(\mathbf{n}, \mathbf{r}^*_{log}; \mathbf{n}, \mathbf{r}^*_{log})$ that under model $\boldsymbol{\psi}_{log}$. In other words, according to (9.9), the lack-of-fit of model $\boldsymbol{\psi}_{log}$ provided model $\boldsymbol{\psi}_{logit}$ true is less than the lack-of-fit of model $\boldsymbol{\psi}_{logit}$ provided model $\boldsymbol{\psi}_{log}$ is true. The robustness derives from the mini-max interpretation: the selected model would have entailed smaller latent deviance had the alternative model been true, than the other way around. In applications, the criterion (9.9) is replaced by its consistent estimator

$$\ell_C(\mathbf{n}, \hat{\mathbf{r}}_{logit}; \mathbf{n}, \hat{\mathbf{r}}_{logit}) - \ell_C(\hat{\boldsymbol{\psi}}_{log}; \mathbf{n}, \hat{\mathbf{r}}_{logit})$$
$$\leq \ell_C(\mathbf{n}, \hat{\mathbf{r}}_{log}; \mathbf{n}, \hat{\mathbf{r}}_{log}) - \ell_C(\hat{\boldsymbol{\psi}}_{logit}; \mathbf{n}, \hat{\mathbf{r}}_{log})$$

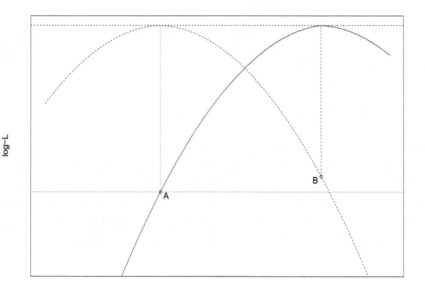

complete data

FIGURE 9.1
Illustration: Maximum $\hat{\ell}_C$ of different models based on pseudo-true data under model A (dashed curve) and B (solid curve); half deviance of model B (vertical dashed line) and A (vertical dotted line); saturated log-likelihood ℓ_C based on different data horizontally aligned (top dashed line).

Figure 9.1 illustrates the LRC heuristically. All the equivalent models satisfying (9.8) yield nil latent deviance given their respective pseudo-true complete data (top horizontal line), where each model is represented by its pseudo-true data placed along the x-axis. The maximum complete-data log-likelihood $\hat{\ell}_C$ is the largest for model A, based on its own pseudo-true complete data. It is represented by the point at which the dashed curve $\hat{\ell}_C$ is tangent to the horizontal line at the top, i.e., nil deviance. The maximum log-likelihood $\hat{\ell}_C$ of the other models decreases as their departure to model A increases, giving rise to the dashed curve of $\hat{\ell}_C$. The maximum log-likelihood of a competing model B is marked by the circle on this curve. Similarly under model B. By the latent LRC one now compares the lack-of-fit of either model under the premise of the competing model, which are represented by the vertical lines: here, model B with the smaller drop (i.e., smaller latent deviance) is deemed more robust against model A from the mini-max point of view.

Additional simulations are recommended to complement the approach. Confidence may be strengthened and more evidence gathered. Basically, what is needed is to choose a plausible outcome subspace that contains the estimated pseudo-true complete-data under the competing models, and to repeatedly fit the models to the different complete-data in the chosen subspace. The plausibility of the latent LRC can be assessed based on the simulations, including relevant bias, mean squared error, etc. of the target prediction. Again, the validity of the simulation results derives from the convergence towards the pseudo-true parameter values and the corresponding pseudo-true data.

9.4.2 Illustration

Let us illustrate the approach outlined above using the synthetic data in Table 9.3. There are 150 enumeration records in A_1, 300 in A_2 and 900 in A_{12}, i.e., 1050 in A_{1+} and 1200 in A_{2+} marginally. The hits and errors among the records that are in only of the lists are held fixed by design, which are (y_1, r_1) and (y_2, r_2). But the list error r_{12} among the records in both lists is allowed to vary for illustration purposes. The empirical error rate is $r_{12}/x_{12} = r_{12}/900$ in A_{12}, and $r_{1+}/x_{1+} = (r_{12}+87)/1050$ in A_{1+}, and $r_{2+}/x_{2+} = (r_{12}+222)/1200$ in A_{2+}. The models to be compared are (9.3) and (9.4), both with $\alpha_{12} = 0$.

TABLE 9.3
Synthetic list hits, errors and survey captures, $N\gamma_A = y_A$.

ω	Capture n_ω	Hit y_ω	Error r_ω	List Size x_ω
$\{1\}$	63γ	63	87	150
$\{2\}$	78γ	78	222	300
$\{1,2\}$	$(900-r_{12})\gamma$	$900-r_{12}$	r_{12}	900
$\emptyset$	$(y_A/\gamma_A - y_A)\gamma$			

Consider first fitting the two models directly based on **r**, which is summarised in Figure 9.2 and comparable to the scheme depicted in Figure 9.1. We notice that the model (9.3) has the best fit at $r_{12} = 18$, where

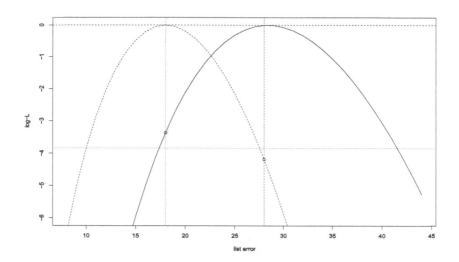

FIGURE 9.2
Fitting log and logit models with $\alpha_{12} = 0$ for $8 \leq r_{12} \leq 44$ in Table 9.3. Maximum log-likelihood of log models (dashed curve) and logit models (solid curve). Critical region of LRT at 5%-level (horizontal dotted line).

$(\hat{\theta}_{12}, \hat{\theta}_{1+}, \hat{\theta}_{2+}) = (0.02, 0.1, 0.2)$ under the PCI-assumption (9.1). Since the fitted error probabilities are exactly equal to the empirical error rates, i.e., $(18 + 87)/1050 = \hat{\theta}_{12+}$, $(18 + 222)/1200 = \hat{\theta}_{2+}$ and $18/900 = \hat{\theta}_{12}$, $r_{12}^* = 18$ is the pseudo-true data of model (9.3). The fit of log model (9.3) worsens as r_{12} moves away from r_{12}^*. However, for $10 \leq r_{12} \leq 27$, it cannot be rejected by the LRT at the 5%-level based on the χ_1^2-distribution with 1 degree of freedom – the critical region is marked by the dotted horizontal line in Figure 9.2.

Similarly, the pseudo-true data of logit model (9.4) is $r_{12}^* = 28.4$, yielding $(\hat{\theta}_{12+}, \hat{\theta}_{1+}, \hat{\theta}_{2+}) = (0.0315, 0.110, 0.209)$, which are equal to the corresponding empirical error rates. For $18 \leq r_{12} \leq 41$, the model cannot be rejected by the LRT at the 5%-level. Thus, the two models complement each other, and together they accommodate a wider range of data than either on its own. In this case, at the 5%-level, the two models can be fitted to $10 \leq r_{12} \leq 41$.

Between the two, the log model fits better than the logit model for $r_{12} \leq 22$ but worse for $r_{12} \geq 23$. The circles in Figure 9.2 mark the LRT statistic of each model based on the pseudo-true data of the other model. Measured in terms of the Kullback-Leibler divergence, the lack-of-fit of the log model when the logit model is true is slightly greater than the other way around.

Having thus demonstrated the rationale of the LRC, we now illustrate how it works when the complete-data LRT statistic is unobserved due to the missing data. For a simulation study in advance, we let the survey catch

TABLE 9.4
Estimates $(\hat{N}, \hat{r}_{12})$ under log model (9.3) and logit model (9.4) with $\alpha_{12} = 0$.
Latent LRC selection marked †. Condition (9.8) holds unless marked *.

	$\gamma_A = 0.95, \gamma = 0.95$			
	$(N, r_{12}, n_A + n_\emptyset)$			
Link	$(1087, 8, 1032)$	$(1077, 18, 1023)$	$(1066, 28, 1012)$	$(1056, 38, 1003)$
log	$(1074, 18.2)^\dagger$	$(1077, 18.0)^\dagger$	$(1078, 17.7)$	$(1081, 17.5)$
logit	$(1061, 29.4)$	$(1063, 28.9)$	$(1065, 28.3)^\dagger$	$(1068, 27.8)^\dagger$
	$\gamma_A = 0.80, \gamma = 0.95$			
	$(N, r_{12}, n_A + n_\emptyset)$			
Link	$(1291, 8, 1226)$	$(1279, 18, 1215)$	$(1266, 28, 1202)$	$(1254, 38, 1191)$
log	$(1276, 18.2)^\dagger$	$(1279, 18.0)^\dagger$	$(1281, 17.7)$	$(1283, 17.5)$
logit	$(1260, 29.4)$	$(1263, 28.9)$	$(1265, 28.3)^\dagger$	$(1268, 27.8)^\dagger$
	$\gamma_A = 0.90, \gamma = 0.98$			
	$(N, r_{12}, n_A + n_\emptyset)$			
Link	$(1148, 8, 1125)$	$(1137, 18, 1114)$	$(1126, 28, 1104)$	$(1114, 38, 1092)$
log	$(1135, 18.2)^\dagger$	$(1137, 18.0)^\dagger$	$(1139, 17.7)$	$(1141, 17.5)$
logit	$(1125, 26.0)^*$	$(1123, 28.9)$	$(1125, 28.4)^\dagger$	$(1127, 27.8)^\dagger$

probability γ and the population size N (set *via* $\gamma_A = y_A/N$) vary in addition to r_{12}. All the list domain sizes x remain fixed as before. The survey captures $(\mathbf{n}, n_\emptyset)$ in Table 9.3 are set to their expected values as follows:

$$(x_1, x_2, x_{12}) \xrightarrow{r_{12}} (y_{12}, y_A) \xrightarrow{\gamma_A} (N, N_\emptyset) \xrightarrow{\gamma} (n_1, n_2, n_{12}, n_\emptyset)$$

Table 9.4 shows the results of a number the choices of $(r_{12}, \gamma_A, \gamma)$, with the corresponding population size N and the total survey captures $n_A + n_\emptyset$. Notice that we allow for greater variation of γ_A than γ in Table 9.4, assuming possibly a relatively higher survey catch probability. The estimates $\hat{N}$ are given under the two models (9.3) and (9.4), respectively, as well as the estimates $\hat{r}_{12} = E(r_{12}|n_{12}; \hat{\psi})$ where $\hat{\psi}$ denotes the corresponding MLE. The result selected by the latent LRC is marked † for each simulation setting.

Firstly, we notice that the MLE of r_{12} centres around the pseudo-true value established in Figure 9.2 earlier, where $\hat{r}_{12}$ is about $r_{12}^* = 18$ under the log model and about $r_{12}^* = 28$ under the logit model, across a range of true data and parameter values. The basis of the latent LRC holds well in this case. Notice also that the pseudo-true data r_{12}^* differs from true data r_{12}, except when the model is true and $r_{12} = r_{12}^*$.

Next, the estimates selected by the latent LRC are always better than the ones not selected. The mini-max criterion works as intended across the settings considered here, which is the range of r_{12} where at least one of the two models cannot be rejected at the 5%-level based on the true data. Indeed, more granular simulations (omitted here) confirm that the turning point is between 24 and 25, where the latent LRC selects the log model for smaller

r_{12} and the logit model for larger r_{12}. In comparison, the turning point of the LRT selection directly based on the complete data r_{12} is between 22 and 23, as can be seen in Figure 9.2. The latent LRC in the presence of missing data thus matches the selection according to the LRT based on the complete data. Finally, the latent LRC selected error-adjusted estimator $\hat{N}$ is always better than the naive survey count $n_A + n_\emptyset$, even when the survey catch probability γ is as high as 0.98 and its relative bias is only 2%.

9.5   Homelessness data in the Netherlands

9.5.1   Data and previous study

Coumans et al. (2017) estimate homelessness in the Netherlands based on capture-recapture data, under the standard log-linear models which assume that erroneous enumeration is absent. The target population is the *roofless* persons in the Netherlands, which constitute a type of homelessness from a broader perspective. See Coumans et al. (2017) for a discussion regarding the relevant definitions. The capture-recapture data are obtained from three administrative registers:

- A list of persons extracted from the Dutch population register (GBA, or G), who has an address of a low threshold service for the homeless, i.e., persons who are registered at an institute which hosts homeless people.

- A list of persons in a register of social benefit (WWB, or W), who according to the register do not have a permanent place of residence.

- A list of persons who are homeless according to the National Alcohol and Drugs Information System (LADIS, or Z).

The reference date is January 1, 2009 in all the three sources. The three-way cross-classified counts are given in Table 9.5. Altogether, there are 5169 enumerated persons. As explained by Coumans et al. (2017), the study concentrates on people between 18 and 65 years old, because there are almost no younger or older people in the three available registers. This reflects the Dutch policy on aid agencies for the homeless, where youngsters and older people are taken care of differently in the social and welfare system. Moreover, people who are not included in the GBA or LADIS register, are still included if they are found in the WWB register, because social benefit is only provided under strict conditions and the WWB register is regarded as trustworthy. As can be seen in Table 9.5, more than half of the people are only enumerated in the WWB register, so this is critical to the analysis. Finally, addresses that provide more than service to homeless people are excluded from the sources, in order to avoid spurious enumeration of out-of-scope individuals.

TABLE 9.5

Cross-classified counts in GBA, WWB and LADIS registers.

(GBA, WWB, LADIS)	Count	(GBA, WWB, LADIS)	Count
$(1, 1, 1)$	30	$(0, 1, 1)$	175
$(1, 1, 0)$	495	$(0, 1, 0)$	2792
$(1, 0, 1)$	24	$(0, 0, 1)$	654
$(1, 0, 0)$	999	$(0, 0, 0)$	

Various log-linear model estimates of the number of homeless people missing by *all* the registers, i.e., m_0 in the notation of this paper, can be obtained using either two of the three registers or all the three registers. However, the estimates range widely from 5635 to 27223. Many of them fit the data either exactly or nearly so. Model selection appears inclusive based on a formal procedure such as the LRT or the Akaike Information Criterion (AIC).

Fortunately, the authors have access to 4 additional covariates: gender, age (3 groups), place of living (2 categories) and country of origin (3 categories). This allows them to construct multi-way incomplete contingency tables, with many degrees of freedom left for model selection. Three best models are reported, obtained by two different search strategies. In particular, (a) these yield similar estimates of the number of people missing by all the three registers, which is around 12589, (b) all the models include the interactions (GBA, WWB) and (GBA, LADIS) but exclude the two-way interaction between WWB and LADIS. See Coumans et al. (2017) for details.

9.5.2 Analysis allowing for erroneous enumeration

With respect to (b) above, one can fit the incomplete log-linear model where $i \in W$ and $i \in Z$ are conditionally independent given $i \in G$, using *only* the data in Table 9.5. This yields $\hat{m}_0 = 10434$ and the deviance 10.98. While $\hat{m}_0$ is plausible compared to the best estimates with additional covariates, the p-value of the model is only about 0.001 based on one degree of freedom. To understand this, notice that the conditional independence implies

$$\Pr(i \in Z | i \in W, i \in G) = \Pr(i \in Z | i \notin W, i \in G)$$
$$\Pr(i \in Z | i \in W, i \notin G) = \Pr(i \in Z | i \notin W, i \notin G) = \pi_0$$

The second equation yields the estimate $\hat{m}_0 = 654(1 - \hat{\pi}_0)/\hat{\pi}_0$, where

$$\hat{\pi}_0 = \widehat{\Pr}(i \in Z | i \in W, i \notin G) = 175/(175 + 2792) \approx 0.059$$

depends only on the counts that do not involve the GBA register. Meanwhile, the first equation does not fit the data well, where the estimates of the left- and right-hand sides are given by, respectively,

$$\widehat{\Pr}(i \in Z | i \in W, i \in G) = 30/(30 + 495) \approx 0.057$$
$$\widehat{\Pr}(i \in Z | i \notin W, i \in G) = 24/(24 + 999) \approx 0.023$$

The two differ quite much, which results in the large deviance. However, the ostensible low catch rate $\widehat{\Pr}(i \in Z | i \notin W, i \in G)$ could be caused by erroneous enumeration in the GBA-count $n_{100} = 999$, which would not affect the other two catch rates above, as long as erroneous enumeration is negligible in the other two registers. This is not unthinkable, due to the incentive and possibility for ineligible persons to 'keep' their places at the institutes that provide service without stringent control, i.e., the source of GBA register.

TABLE 9.6
Alternative estimates $\hat{m}_0$ and associated deviances.

Model	$n_{100} = 999$		$n_{100} = 396$	
	$\hat{m}_0$	Deviance	$\hat{m}_0$	Deviance
[GW][GZ][WZ]	26322	0	10434	0
[GW][WZ]	27223	0.03	10791	0.03
[GW][GZ]	10434	10.98	10434	0
[GW]	12240	24.00	10518	0.4

Now, it is easily verified that, by changing the count n_{100} from 999 to 396 but keeping all the other counts in Table 9.5, one would retain the estimate $\hat{m}_0 = 10434$ while reducing the deviance of the model [GW][GZ] to zero. In other words, this is the pseudo-true data under this model. In Table 9.6 we compare the estimates of m_0 and the associated deviances of the four best-fitting models based on the original data, where $n_{100} = 999$, to those based on the stipulated pseudo-true data where n_{100} is changed to 396 (i.e., allowing for erroneous enumeration in the GBA). The change is seen to improve the goodness-of-fit, as well as stabilising the estimates.

Exploratory analysis above indicates that there is room for considering models that allow for erroneous enumeration. However, without the additional covariates, the models that can be introduced are rather restrictive. The analysis below does not aim to produce alternative estimates of homeless people to the study of Coumans et al. (2017), but rather it serves to demonstrate the potential usefulness of the modelling approach to erroneous list data.

Treating any of the three registers as the enumeration survey and allowing for erroneous enumeration in the other two registers, i.e., $K = 2$, the largest unsaturated model is given either by (9.3) or (9.4) with $\alpha_{12} = 0$. This gives us 6 possible models, all with zero degree of freedom. In Table 9.7 we summarise the deviance, the parameter estimates and the estimated missing count m_0.

Firstly, the results are abnormal when the WWB register is treated as the enumeration survey, where there are big differences between the log and logit models. The logit model does not fit the data well despite zero degree of freedom; the log model has a much smaller deviance but yields implausible estimates of the error probabilities θ_+ and the missing count m_0. On further examination it becomes clear that the log-likelihood is very flat under the

TABLE 9.7

Results of log and logit models.

Model	Deviance	$\hat{\theta}_{1+}$	$\hat{\theta}_{2+}$	$\hat{\theta}_{12+}$	$\hat{\gamma}$	$\hat{m}_\emptyset$	$n_\emptyset$
$S = \mathrm{GBA}, A_{1+} = \mathrm{WWB}, A_{2+} = \mathrm{LADIS}$							
log	0.03	0.007	0.589	0.004	0.151	5597	999
logit	0.01	0.030	0.602	0.045	0.155	5447	999
$S = \mathrm{WWB}, A_{1+} = \mathrm{GBA}, A_{2+} = \mathrm{LADIS}$							
log	0.98	0.657	0.767	0.504	0.987	38	2792
logit	9.83	0.129	0.425	0.099	0.388	4409	2792
$S = \mathrm{LADIS}, A_{1+} = \mathrm{GBA}, A_{2+} = \mathrm{WWB}$							
log	0.03	0.399	0.005	0.002	0.059	10434	654
logit	0.03	0.401	0.009	0.006	0.059	10383	654

log model here, so that the estimators are highly uncertain, despite the small deviance. The reason is that there is only a very small overlap between the GBA and LADIS registers, i.e., the list domain A_{12} with only 54 units, while most of the records are in either of the two registers, with $x_1 = 1494$ and $x_2 = 829$. The situation is quite different to that considered in Section 9.4.2, where most of the units are in the overlapping list domain A_{12}.

There is little difference between the log and logit models when either the GBA or LADIS register is treated as the enumeration survey. Under both sets of the models, the WWB register is estimated to have low list error probability, where three out of four estimates of θ_{W+} are below one percent and even lower for θ_{WZ+} or θ_{GW+}. This confirms the trust of Coumans et al. (2017) in this register. However, while both sets of models have almost nil deviance, they give very different estimates of $m_\emptyset$.

The estimate $\hat{m}_\emptyset$ under the log model with $S = \mathrm{LADIS}$ is the same as that under the log-linear model [GW][GZ] above. Indeed, the estimated error probability is $\hat{\theta}_G = 0.602$ in the GBA-only domain, i.e., $\hat{r}_G = 999\hat{\theta}_G = 602$ and $\hat{y}_G = 999 - 602 = 397$, which is about the same as the previously stipulated pseudo-true data value in Table 9.6. This shows that the formally developed list-error models can provide a more theoretical approach in place of the kind of exploratory analysis above.

The estimates $\hat{m}_\emptyset$ are much lower at around 5500 under the log and logit models with $S = \mathrm{GBA}$. Moreover, the LADIS register is estimated to have a high error probability, where $\hat{\theta}_Z$ is about 0.77. We do not know whether this can be plausible at all, in light of the source of LADIS register. Had we access to the additional covariates, we could have fitted more detailed models with enough degrees of freedom. It is quite possible that model selection can then be settled based on a formal procedure such as the LRT or the AIC, just like in the study of Coumans et al. (2017). Lacking this option, however, we shall assess the two sets of models using the latent LRC.

To construct the pseudo-true complete-data under a given model, one calculates the estimated conditional expectation of the counts by the four-way cross-classification $G \times W \times Z \times U$. There are 5 structure zero cells under the models here. For example, suppose S = LADIS, the four cells $(G, W, Z, U) = (-, -, 1, 0)$ have zero counts because it is not possible to have survey captures that do not belong to the target population, now that the enumeration survey is assumed to only suffer from under-enumeration. The fifth structural zero cell is $(G, W, Z, U) = (0, 0, 0, 0)$ by definition.

The zero cells are the same for a competing model that assumes the same enumeration survey, when calculating the estimated probabilities of the complete-data. But they will differ when the competing model assumes a different enumeration survey. For example, suppose the pseudo-true data are calculated assuming S = LADIS, with the 5 structural zero cells given above. Let the competing model assume S = GBA. The two cells $(G, W, Z, U) = (1, -, 1, 0)$ remain zero, because the survey cannot have any captures outside the target population, and the cell $(0, 0, 0, 0)$ remains zero as always. However, the other two structural zero cells will differ. The two cells $(G, W, Z, U) = (0, -, 1, 0)$ which used to be zero are no longer zero, while the two cells $(G, W, Z, U) = (1, -, 0, 0)$ which used to be non-zero are now zero. This is a reason why the latent deviance tends to be much bigger when the competing models assume different enumeration surveys than otherwise.

TABLE 9.8
LRC selection marked †

	Model of pseudo-true data	Model fitted	Latent deviance
(I)	(S = GBA, log)	(S = GBA, logit)	138.04
	(S = GBA, logit)	(S = GBA, log)†	137.59
(II)	(S = LADIS, log)	(S = LADIS, logit)	14.22
	(S = LADIS, logit)	(S = LADIS, log)†	12.73
(III)	(S = GBA, log)	(S = LADIS, log)†	10208.12
	(S = LADIS, log)	(S = GBA, log)	19247.94
(IV)	(S = GBA, logit)	(S = LADIS, log)†	10599.38
	(S = LADIS, log)	(S = GBA, logit)	19476.44
(V)	(S = GBA, log)	(S = LADIS, logit)†	10225.69
	(S = LADIS, logit)	(S = GBA, log)	19310.55
(VI)	(S = GBA, logit)	(S = LADIS, logit)†	10524.25
	(S = LADIS, logit)	(S = GBA, logit)	19500.54

Table 9.8 provides a summary of the comparisons between the two sets of models with S = GBA or LADIS. The first two comparisons (I) and (II) are between the log and logit models, assuming the same enumeration survey. The differences in the deviances are small, but the log model is favoured by the latent LRC in both cases. The next four comparisons (III) - (VI) are between models assuming different enumeration surveys. The latent deviances are much larger, as well as the differences between them. Nevertheless, the

latent deviances are approximately of the same magnitude when the pseudo-true data are calculated assuming the same enumeration survey, whether it is the GBA or LADIS register. This suggests that the assumption about the enumeration survey affects the results more than the choice of the link function. In fact, the model assuming $S = $ LADIS is favoured by the latent LRC in all the cases, regardless of the choice of link functions, which yields the estimates $\hat{m}_0$ that are more plausible in light of the results obtained by Coumans et al. (2017) with the help of additional covariates. Thus, the analysis provides an empirical example that model selection by the latent LRC can be successful in the presence of missing data, where the standard formal procedures such as the LRT and the AIC are not useful.

Bibliography

[1] BÖHNING, D., VAN DER HEIJDEN, P.G.M. and BUNGE, J. (2017). *Capture-Recapture Methods for the Social and Medical Sciences*. Chapman & Hall/CRC.

[2] CORMACK, R.M. (1989). Log-linear models for capture-recapture. *Biometrics* **45** 395–413.

[3] COUMANS, A.M., CRUYFF, M., VAN DER HEIJDEN, P.G.M., WOLF, J. and SCHMEETS, H. (2017). Estimating homelessness in the Netherlands using a capture-recapture approach. *Social Indicators Research* **130** pp. 189–212.

[4] DEMPSTER, A.P., LAIRD, N.M. and RUBIN, D.B. (1977). Maximum likelihood from incomplete data via the EM algorithm (with discussion). *Journal of the Royal Statistical Society, Series B* **39** 1–38.

[5] FIENBERG, S.E. (1972). The multiple recapture census for closed populations and incomplete 2^k contingency tables. *Biometrika* **59** 409–439.

[6] IWGDMF - INTERNATIONAL WORKING GROUP FOR DISEASE MONITORING AND FORECASTING (1995). Capture-recapture and multiple-record systems estimation I: History and theoretical development. *American Journal of Epidemiology* **142** 1047–1058.

[7] IWGDMF - INTERNATIONAL WORKING GROUP FOR DISEASE MONITORING AND FORECASTING (1995). Capture-recapture and multiple-record systems estimation 2: Applications. *American Journal of Epidemiology* **142** 1059–1068.

[8] NIREL, R. and GLICKMAN, H. (2009). Sample surveys and censuses. In *Sample Surveys: Design, Methods and Applications, Vol 29A (eds. D. Pfeffermann and C.R. Rao)*, Chapter 21, pp. 539–565.

[9] VUONG, Q.H. (1989). Likelihood ratio tests for model selection and non-nested hypotheses. *Econometrica* **57** 307–333.

[10] WOLTER, K. (1986). Some coverage error models for census data. *Journal of the American Statistical Association* **81** 338–346.

[11] WU, C.F.J. (1983). On the convergence properties of the EM algorithm. *Annals of Statistics* **11** 95–103.

[12] ZHANG, L.-C. (1996). *Latent Likelihood Analysis of Binary Panel Data*. Unpublished Dr. Scient. Thesis.

[13] ZHANG, L.-C. (2015). On modelling register coverage errors. *Journal of Official Statistics* **31** 381–396.

[14] ZHANG, L.-C. and DUNNE, J. (2017). *Trimmed Dual System Estimation*. In *Capture-Recapture Methods for the Social and Medical Sciences (eds. D. Böhning, P.G.M. van der Heijden and J. Bunge)*, Chapter 17, pp. 239–260. Chapman & Hall/CRC.

10

Sampling design and analysis using geo-referenced data

Danila Filipponi, Federica Piersimoni
Istat, National Institute of Statistics, Rome, Italy

Roberto Benedetti
Department of Economic Studies, University "G. D'Annunzio" of Chieti-Pescara, Pescara, Italy

Maria Michela Dickson, Giuseppe Espa, Diego Giuliani
Department of Economics and Management, University of Trento, Trento, Italy

CONTENTS

10.1 Introduction

In statistics, the units of a population under observation could be studied under a double point of view, simply considering the information about them under an a-spatial perspective or considering where they are located on a territory and in what way they interact with other units positioned in the same space. Therefore, an important question to answer concerns where a phenomenon happens, for example, where a city is located or where an establishment has its plant.

Classical statistical studies typically assume that data are one-dimensional, but it is not true in the context of spatial statistics. In this case, instead, data are viewed as points in, at least, a two-dimensional Cartesian coordinate system [38] identifying the locations of units on a specific territory, in which and from which they carry out actions and interactions [5]. Spatial statistics includes many techniques to study spatial phenomena, based on different analytic approaches and applied in many different fields, such as ecology, economics, epidemiology, image analysis and more [23, 25, 50, 54]. One of the contexts in which spatial aspects of units are very important is sampling theory. Here, the knowledge about locations of population units in a geographic space can be very helpful in studying phenomena characterised by spatial dependence and heterogeneity [2, 3].

Spatial sampling is a prosperous field of research that in recent years has produced several new methodologies. Basic sampling schemes include random, clustered and systematic methods, which take into account spatial information in the phase of sampling design. These methods can be applied according to different resolutions of the spatial information (i.e., cities, urban areas, neighbourhoods, regions) where the finest level consists of the individual point locations of the units, as identified by the spatial coordinates (namely longitude and latitude). Geographic coordinates are identified from information collected through Geographical Information Systems (GIS), which obtain, register, analyse and visualize geo-referenced data [13]. Moreover, the availability of spatial information about units leads to their use as an important auxiliary variable, which could be used as a strong informational component in, among others, stratification and estimation procedures. As said before, spatial sampling could have several applications in agriculture, environmental studies and, more in general, social science [11, 21, 28, 30, 44, among others]. In this chapter, we focus the attention on geo-referenced administrative data, such as, in particular, population and business registers in Italy. These archives could be affected by problems of various nature, i.e., the choice about the data model, the quality of the interpolation process and of the input data address, the accuracy of reference data and of the matching algorithm used. In order to cope with these problems and to obtain meaningful results, the geo-coding process must respect some standards of quality about completeness of registers, locational accuracy of the census points and repeatability of the implemented study.

In reality, registers are inevitably affected by errors, which can lead to imprecise sample surveys. In this chapter, we concentrate the attention on problems related to completeness and locational accuracy of Italian geo-referenced registers and on consequences in sampling methods applied on them. In Section 10.2 are presented the traditional problems in geo-coding procedures and the deriving errors in registers. Section 10.3 contains a detailed explanation of spatial sampling methods for geo-referenced registers. Section 10.4 and Section 10.5 illustrate two possible applications of spatial sampling in presence of coverage errors, as a consequence of failed completeness of archives, and sampling by imprecise data, as a consequence of locational errors. Simulation studies are conducted for both applications. In Section 10.6 some conclusions and ideas for further studies are outlined.

10.2   Geo-referenced data and potential locational errors

Geo-coding of registers is a procedure consisting in assigning a (x, y) pair of coordinates to a specific place or unit by means of a comparison between the location detected by a GPS device and the address contained in a referenced system of data[1]. Therefore, after the location of a unit is identified, it is necessary to verify that it corresponds to its actual street address. Therefore, each detected location is compared with a set of possible addresses candidate to match it and the best one is chosen as a point on a map. This procedure starts with an analysis of input addresses into address components (e.g., street names and street types); then values are standardized and they are matched with each part of the original address element. So, it is assigned a score to any potential candidate for a position and the choice is guided by a match score.

Geo-coding techniques are based on deterministic or probabilistic linkages. The deterministic approach is based on a comparison between each address with reference elements and a score is given on the basis of how good is the match between them. Then, for each element, a set of weights is used to compute an overall score, which is compared with the minimum level score given by the user. Instead, in the probabilistic approach (*fuzzy matching*) a set of weights is computed for each element, on the basis of the estimated capability to correctly identify a match or a missed match. These weights are then used to compute the probability that two records are referred to the same point [37, 55, 58, 61].

In geo-coding procedures, the most important requirement is the availability of correct reference data and a robust model for checking the addresses, in order to organize reference data in a logical way. There are numerous models to verify addresses, which can be divided in groups, based on the provided geo-coded information. The first group (*geographic unit models*) consists of models that use administrative information such as postal codes, provinces, regions or other geographic borders. In this geo-coding process, the position is assigned to a polygon, which represents the geographical

---

[1]Geo-coding of statistical units must not be confused with the official nomenclatures of territorial units (NUTS, Nomenclature des Unités Territoriales Statistiques). This geographical nomenclature subdivides the economic territory of European Union in areas, composed by NUTS 1 (national level), NUTS 2 (regions) and NUTS 3 (districts or provinces). The NUTS is governed by Regulation (EC) No 1059/2003 of the European Parliament and of the Council of 26 May 2003 on the establishment of a common classification of territorial units for statistics.

unit. Here the goodness of results strictly depends on the size of the unit; such that, if it is very big, micro-level studies cannot be conducted. However, in these cases, a second group of models (*address data models*) can solve the problems because they are based on the use of street networks or address points (for technical details, see [15]).

The completeness is the percentage of records that can be exactly geo-coded, according to the matching rate, which can vary depending on many factors. The matching rate increases if there is an improvement in the quality of address files and/or geographic reference files. There is not a common idea about the acceptable level of a matching rate. Ratcliffe [41] argues that a matching rate of at least 85% is necessary. Clearly, an incomplete geo-coding is a source of bias because populations with different proportions of geo-coded and not geo-coded addresses can lead to imprecise analysis of data. Concerning locational accuracy, it is defined as how close a geo-coded point is to the real position of its address in a geographic space. Several components can contribute to errors in locational accuracy, i.e., points assigned to a wrong street segment, due to errors in input address fields or street databases; points assigned to a correct segment of a street, but the entire segment contained in the reference database is incorrect; the interpolated assignment of an address alongside the street segment may not coincide with the correct location of the address, due sometimes to a lacking correspondence between linear and effective numbering scheme of addresses on a segment. Consequently, the sensitivity of geo-coded results with respect to variations in reference data and matching algorithms can undermine the repeatability of spatial studies.

It is possible to track some similarities between traditional record linkage techniques and these three requirements. As in record linkage, also problems in matching between reference data, administrative data and geo-coded data can lead to a difficult use of this information. In the rest of the chapter, we will deepen effects of incompleteness and locational errors on spatial studies.

10.3   A brief review of spatially balanced sampling methods

Typically, near units have similar characteristics, depending on the intrinsic nature of data. It is advisable in sampling from a finite population to select units, which are not adjacent. The idea behind sampling designs is to exclude contiguous units to be selected in the same sample. Literature in this field is extensive and the following cited works do not compose an exhaustive review. Hedayat et al. [32] and [33] first introduced a sampling method without replacement based on controlling second-order inclusion probabilities, which assigned value zero for contiguous units and constant for non-contiguous ones. Moreover, Stufken [51] enhances balanced sampling avoiding adjacent units. He generalized this concept by excluding in the selection units with distance between each other less than, or equal to, a specified number m. Stufken [52] introduced also polygonal designs as a group of designs equivalent to balanced sampling avoiding adjacent units. Following this field of research, two important contributions are given by Mandal et al. [39] and Wright and Stufken [57]. In the first one, a new family of distance balanced sampling methods is presented, in which the second-order inclusion probabilities are managed as a non-decreasing distance function between units. The second one presents another family of techniques for the exclusion of adjacent units in presence of different population types.

The methods listed before, consider the distance between two units in the population of interest and find solutions to exclude in the selection units that are close in terms of a defined distance function. Moreover, others methodologies try to consider distance under a different point of view, continuing to avoid the selection of near units. We cite, among others, the *Dependent Areal Units Sequential Technique* (DUST) [4] and the *Generalized Random Tesselation Stratified* (GRTS) method [49]. The first is a sequential technique, which incorporates spatial correlation in the sample selection and, basing on it, sampling units have the same probability to be drawn that increases when their distance from the sampled areas increases [4]. Instead, GRTS is a method that uses a function that maps a space from two dimensions to one dimension, trying to preserve the spatial order of units. To reach this aim, GRTS divides the space into cells and applies a mapping function that assigns an order to the units. Then, a sample is selected in one dimension, by using systematic πps, and then it is mapped back in two dimensions. The method could be implemented for point, linear and areal frames and it has been the main spatial method used in environmental studies for many years [49].

A more recent and advanced field of research uses a distance function directly in complex algorithms, which select units trying to draw samples well spread in the space, as explained below. This is the case of Local Pivotal Methods [28], Spatially Correlated Poisson Sampling [27] and Local Cube Method [29]. These methodologies will be reviewed in the following subsections (for details, see [8, 9]). Special consideration is needed for another methodology, which is the Balanced Sampling through the Cube Method [19], that will be described and used in the applications in its spatial version. We have omitted the dissertation about more recent methods [7, 22] and about variance estimation [6, 48].

All methods described below may be implemented with the R packages "sampling" [53] and "Balanced Sampling" [26].

10.3.1 Local pivotal methods

Local Pivotal Methods are two fast and flexible algorithms proposed by Grafstrom et al. [28], based on the Simple Pivotal Method [18] extended to consider the spatial aspect. Under Pivotal Method, an equal or unequal probability sample is drawn in N steps, in which, at each step, the inclusion probabilities are updated for two units. If the updated inclusion probabilities π_i' are equals to 0 or 1, the unit i is *finished* and it may not be chosen again. The updating procedure is repeated until all units are visited and chosen for the sample. In their methods, Grafstrom et al. [28] update the inclusion probabilities according to the updating rule by Deville and Tille [18] but for two close units. To choose two close units i and j and to compute distance between them, they proposed two different methods (LPM1 and LPM2). The LPM1, which has more balance, randomly chooses the first unit i and then the nearest neighbor unit j (if two or more units have the same distance to i, the method randomly chosen between them). If j is not the nearest neighbor of i, the method restarts. When the algorithm has visited all units, then it stops. LPM2 works similarly to LPM1, but the inclusion probabilities are directly updated with the updating rule of the pivotal method. The expected number of computations needed to select a sample is proportional to N^2 instead of N^3 as for LPM2. Both algorithms draw spatially balanced samples, which could be evaluated following the Voronoi polygons approach proposed by Stevens and Olsen [49].

10.3.2   Spatially correlated Poisson sampling

Spatially Correlated Poisson Sampling (SCPS, [27]) is a technique that selects spatially balanced samples, such as with locations of units that are well distributed over the population. The method is based on a modification of the Correlated Poisson Sampling [10] but taking into account spatial information. The method visits each unit one time and decides whether or not it should be sampled, with the intent to create positive (negative) correlation between the inclusion indicators for units that are far (close) in distance. If correlation is negative, then units close in the space are rarely drawn in the same sample, and vice versa, thus producing samples that are spread in the space. SCPS uses the probability function of [10], which can be written as

$$Pr(I = x) = \prod_{i=1}^{N} \left(\pi_i^{(i-1)} \right)^{x_i} \left(1 - \pi_i^{(i-1)} \right)^{1-x_i}, x \in \{0, 1\}^N$$

For unit i, the probability $\pi_1^{(0)} = \pi_1$ set $I_1 = 1$ and otherwise $I_1 = 0$. At each step of the method, the inclusion probabilities for the remaining units are updated, according to a specific rule, described in Bondesson and Thorburn [10]. The method uses a system of weights, which vary for each unit to the previous one. Grafstrom [27] introduces a known distance between units into Correlated Poisson Sampling method. The distance among units determines a system of weights that lead to spatially balanced samples.

The distance function $d(i, j)$ between units i and j may be the Euclidean distance or another general distance measure. The function gives an order to the units and put them in a list, from which samples are drawn. Grafstrom [27] proposed two different strategies for choosing weights: by maximal weights and by Gaussian preliminary weights. The samples will be well spread over the population and the efficiency of the method is independent of the ordering of units.

10.3.3 Balanced sampling through the cube method

Balanced sampling, as it is well known, is a sampling methodology defined by the property of the Horvitz-Thompson estimators [34] on a set of balancing variables if the estimated totals of these variables are equal to the population totals. A large family of methods to select balanced samples is available in literature, which includes, among others, those proposed by [17, 31, 43, 59].

The most known method to select approximately balanced samples, with equal or unequal inclusion indicators and any number of variables, is the Balanced Sampling Cube method (BSC) [19]. It allows the selection of exactly balanced samples, if it is possible to find them, or otherwise approximately balanced samples. It is based on a geometric representation of sampling design. The Cube method works in two phases. First, during the so-called *Flight phase*, a random walk starts and stays in the intersection area between the cube and its subspace, stopping only when reaching a vertex of the subspace. In this phase, the aim is to exactly satisfy the balancing constraints, having almost all the inclusion probabilities equal to zero or one. Then, a *Landing phase* is applied, in which the balancing constraints are relaxed in order to select any sample as near as possible to the constraints subspace.

As said, the Cube method can be applied for any type and number of auxiliary variables. So, in the present work, we use the geographical coordinates of units as auxiliary variables and implement the method by means of the fast algorithm [14].

10.3.4 Local cube method

Grafstrom and Tille [29] proposed the Local Cube Method (LCM), a doubly spatially balanced methodology, which is a combination of LPM2 [28] and the Cube method [19]. This new algorithm draws samples that are balanced on p auxiliary variables and spread in a space. In the first phase, the method applies the LPM2 algorithm to select units, according to the distances between them. The first unit i is randomly selected; then the second selected unit is the nearest to i. So, the mean position it is computed between these units, and the third unit is selected as the nearest to this mean, and so on. This procedure is repeated as many times as the sum of the squares of the distances between cluster units and their means decreases [28]. In this way, LPM2 selects a cluster of near units, wherein LCM imposes the flight phase of Cube method [19]. According to balancing conditions, outcomes are decided for the sample units. The inclusion probabilities are updated at each selected unit locally for the remaining units, in a way that they are small between near units creating negative correlation between sample inclusion indicators [28]. When the flight phase and outcome decision process ends, then the algorithm applies the landing phase of the Cube method.

The result of LCM is the selection of samples that have a good degree of balancing, due to the Cube method, and are well spread in the space, because LPM2 prevents the selection of close units. Development of LCM with respect to its predecessors is that it needs two different vectors, one of auxiliary variables and one of geographic coordinates. It allows to well manage the space component in the procedure of sample selection.

10.4 Spatial sampling for estimation of under-coverage rate

One of the main problems deriving from incompleteness of registers is coverage errors, intended as a gap between sampling frames and population totals. These errors are a kind of non-sampling errors and can lead to biased results in studies that use data coming from not perfectly covered registers.

An administrative register is affected by under-coverage if there are units belonging to the target population not enumerated in the register. On the contrary, it is affected by over-coverage if some of the units enumerated in the register are not in the target population. As an example, if the target population is the population of inhabitants of a country, under-coverage of the administrative population register is usually given by illegal population or by administrative delays in the registration like births and immigration; whereas over-coverage is usually given by administrative delays in the registration like deaths and emigration. One of the main surveys on population conducted by the Italian National Institute of Statistics (ISTAT) is the *Population and Housing Census*. It has a central role in the integrated national statistical system, which may include other censuses, surveys and statistical registers. The census provides, at regular intervals, accurate population counts and information on social, demographic and economic characteristics for small geographical areas or sub-populations. However, traditional censuses are the most difficult and costly data collection activity that ISTAT carries out. Due to this and other considerations, population censuses are usually conducted only once every ten years, so data are available several years after the reference date. In order to reduce costs and statistical burdens and to improve coverage, timely dissemination and frequency of the statistical production, registers and other administrative sources are gradually becoming a practicable alternative to the traditional census in an increasing number of countries. However, statistics about populations based on administrative registers can be obtained for all geographical areas, since registers usually cover the whole target population and contain detailed geographical information from which can be derived all geographic units[2].

In some cases, population totals are difficult to count. So, the so-called *Capture-Recapture models* are traditionally used to deal with population-size estimation. Classical examples include fishes, birds, and other types of animals (for a comprehensive review about animal abundance issues, see [40, 45, 46, 47, 60]). In official statistics context, these models have found application to evaluate census coverage errors [24, 56].

The foundation of capture-recapture models is to estimate the unknown size of a population of interest, by using information collected by multiple lists, related to the same population. Population units are classified according to the lists, in which they are included. If we suppose that J lists are available, a *capture history* vector $s = \{s_1, ..., s_J\}$ identify whether the subject was captured on each of the J lists, with $s_j = 1$ if list j recorded the individual, and $s_j = 0$ otherwise, where $j = 1, ..., J$ index the lists. Data can be regarded as a form of an incomplete 2^J multi-way contingency table, created by summing over individuals with similar capture histories, such that n_s is the number of individuals with history s, where $s = 1, ..., 2^J$ index the capture histories. Then, in the 2^J multi-way contingency table $L = 2^J - 1$ is the number of observable capture histories, where $\sum_{s=1}^{L} n_s = n$ is the number of individuals that have been captured by at least one list, and the history vector $s = \{0, ..., 0\}$ identifies individuals not seen by any list. The goal is to estimate the number of unobserved individuals $N - n$, or likewise of coverage errors of the lists.

To estimate the number of unobserved individuals, a definition of a model is needed that specifies the capture probability (i.e., the probabilities of being observed in the different lists), the dependences among lists, and how the capture probabilities and list dependences vary across individuals. In the simplest experiment, two lists A (population registers or other administrative sources) and B (census or post-enumeration surveys) are available. Each list may be affected by coverage errors, as lack of some population units or counting units not belonging to the population. In Table 10.1, an example is described, where the number of individuals captured by both lists, the number of individuals captured by list A and not by list B, the number of individuals captured by list B and not by list A, and the number of individuals that are neither captured by A nor B are shown.

Capture-recapture analyses for census under-coverage errors typically adopt the assumptions stated in Wolter [56]. In the present work, we treat list A as fixed, so that we assume constant catch rate in the survey (or census), no matching errors, no spurious records (over-coverage) in either list due to duplicates, reporting errors, etc. As a result, a model-based estimator

---

[2]Note that in this study, we refer to units as administrative divisions, i.e., municipalities, enumeration area, map grids of different sizes.

TABLE 10.1

Two-way contingency table for
two list example.

List A		List B		
		In	Out	
	In	n_{11}	n_{10}	n_{1+}
	Out	n_{01}	n_{00}	n_{0+}
		n_{+1}	n_{+0}	n_{++}

of the population size is given by the so-called Dual-System Estimator (DSE).

$$\hat{N} = \frac{n_{1+}n_{+1}}{n_{11}}$$

and the coverage rate estimator is

$$\tau = \frac{n_{11}}{n_{+1}}$$

Usually, a second list is available in the form of area sample survey, also called post-enumeration survey. The survey aims to enumerate the whole population at the sampled locations, usually census tracts.

Defining $n_{i,ab}$ as the number of individuals of the tract i in cell ab, for $a, b = 1, 2, +$, the quantities n_{11} and n_{+1} can to be estimated from the sample as $n_{11} = \sum_i w_i n_{i,11}$ and $n_{+1} = \sum_i w_i n_{i,+1}$, where w_i are sampling weights. Thus, DSE for the whole population is $\hat{N} = \frac{n_{1+}n_{+1}}{n_{11}}$. By using similar considerations, the coverage rate can be estimated by $\tau = \frac{n_{11}}{n_{+1}}$.

Post-enumeration surveys implemented by ISTAT generally assume that a simple random sample of census tracts is selected, without considering the locations of units in the space. Clearly, the method can be easily extended to more complicated sampling designs. The study here aims to investigate the effect of spatial sampling designs in a capture-recapture setting to evaluate under-coverage errors. We consider data coming from 2011 Italian popula-tion register and from 2011 Italian population census. The Italian adminis-trative population register is managed at a local level by around 8100 Ital-ian municipalities, each of them in charge of its territorial. This system of a municipal population register is continuously updated with individual data for administrative purposes. Starting in 2010, ISTAT acquired the municipal population registers through a common data transmission protocol. A web-based application was used to collect and check data transmitted by the mu-nicipalities. The final register is made up of around 61 million individuals and 25 million households. Moreover, each statistical unit has an identifica-tion number and all the addresses are standardized and geo-coded according to the census tract code.

On the point of view of capture-recapture models, population register is employed as List A and population census as List B. Data necessary to esti-mate n_{11} are available thanks to the linkage operation carried out after the census operations, trying to take into account over-coverage errors or matching errors, which can affect the accuracy of the results [20, 62]. Our study is focused on one Italian region, such as Emilia Romagna, characterized by 4.5 million people, according to the population register, and $N = 35585$ census tracts. Figure 10.1 shows the coverage rate τ of the population register for the N census tracts. Bigger dots represent areas with higher coverage rate, where the values of τ are assigned to the centroids of the census tracts. Data exhibit a spatial correlation, with spatial clusters in the center of the region and on the east side.

FIGURE 10.1
Coverage rate in the Emilia Romagna census tracts.

In order to mimic area samples used as List B, we draw samples from the 2011 Italian population census by using spatial sampling methods described in Section 3. In particular, we compare the Local Pivotal Method 1 (LPM1) [28], the Local Pivotal Method 2 (LPM2) [28], the Spatial Correlated Poisson Sampling (SCPS) [27], the Local Cube Method [29] both constrained to a linear trend (LCM1) and to a quadratic trend (LCM2). Moreover, all the sampling designs have been implemented with equal and unequal (proportional to size of register population n_{1+}) inclusion probabilities. In order to compare efficiency of spatial sampling designs, simple random sampling without replacement (SRSWOR) has been implemented as benchmark in equal inclusion probability case and probability proportional to size sampling (PPS) as benchmark in unequal inclusion probability case. All spatial sampling designs have been used to draw 1000 samples from the 2011 Italian population census of size n equal to 350, 1750 and 3500 census tracts. For each sampling design and all sample sizes considered, the Root Mean Square Error has been estimated

$$RMSE(CovRate) = \sqrt{\sum_{nsim} (\tau - \bar{\tau})^2 / nsim}$$

where *nsim* represents the number of Monte Carlo repetitions [42]. In this study *nsim* = 1000. We have evaluated the advantage of including spatial information in the sampling design in terms of relative efficiency gain, defined as

$$RelEff = \frac{RMSE(CovRate)_{spatialdesign}}{RMSE(CovRate)_{a-spatialdesign}}.$$

In addition, the spatial balance of drawn samples has been computed [49]. A Voronoi polygon includes in the sample *s* a unit *i* and all units closer to *i*. Then, the polygon works in the same way for unit *j*. If one unit has the same distance between two or more other units, then these units are included in more than one polygon. By using the variance

$$SBal = \frac{1}{n} \sum_{i \in s} (v_i - 1)^2$$

of the unit $i \in s$, where v_i is the sum of the inclusion probabilities of all units within the polygon, it is possible to measure spatial balance index for a sample [28]. The results about the estimation of parameters and spatial balance are shown in Table 10.2.

TABLE 10.2
Relative efficiency of spatial sampling methods for equal and unequal inclusion probability.

Design	Equal inclusion probability				Design	Unequal inclusion probability			
	n	*N*	*τ*	*SBal*		*n*	*N*	*τ*	*SBal*
LCM2	350	0.908	0.908	0.251	SCPS	350	1.226	1.210	0.624
LCM1	350	0.914	0.914	0.261	LPM2	350	0.880	0.880	0.630
SCPS	350	0.924	0.924	0.275	LPM1	350	0.928	0.929	0.651
LPM1	350	1.001	1.001	0.285	LCM2	350	0.910	0.911	0.719
LPM2	350	0.883	0.883	0.291	LCM1	350	1.131	1.126	0.778
LCM1	1750	0.963	0.963	0.291	LPM1	1750	0.876	0.877	0.657
LCM2	1750	0.939	0.939	0.303	LPM2	1750	1.302	1.296	0.670
LPM1	1750	0.938	0.938	0.318	SCPS	1750	1.036	1.036	0.674
SCPS	1750	0.913	0.913	0.322	LCM2	1750	2.750	2.665	0.767
LPM2	1750	0.930	0.930	0.323	LCM1	1750	2.055	2.023	0.813
LCM2	3500	0.903	0.903	0.313	LPM1	3500	1.121	1.119	0.690
LCM1	3500	0.914	0.914	0.319	LPM2	3500	0.846	0.846	0.697
SCPS	3500	0.889	0.889	0.333	SCPS	3500	0.922	0.922	0.708
LPM1	3500	0.880	0.880	0.333	LCM2	3500	1.251	1.249	0.820
LPM2	3500	0.914	0.914	0.335	LCM1	3500	1.341	1.337	0.825

For all spatial sampling methods with equal inclusion probability, relative efficiency is around 10% both for the estimation of the population total and for the coverage rate. In particular, the best result has been obtained with a sample of 3500 units. Instead, for the spatial sampling methods with unequal inclusion probability, the relative efficiency is lower, because the first order inclusion probabilities are weakly correlated with target variable values. Moreover, we consider the problem of estimating the coverage rate for not planned small geographic areas within the region. Both estimation methods discussed in Section 10.3. are difficult to apply within the small area because there is no sample from List B in the considered area or the sample is too small. Tables 10.3 and 10.4 show the relative efficiency of the estimation of population size N for the nine provinces in Emilia Romagna region.

TABLE 10.3
Relative efficiency of the estimation of population size in small geographical domains with unequal inclusion probability.

Design	n	Piacenza	Parma	Reggio Emilia	Modena	Bologna
SCPS	350	0.219	0.326	0.382	0.304	0.265
SCPS	1750	0.136	0.189	0.226	0.188	0.167
SCPS	3500	0.120	0.162	0.187	0.158	0.148
LPM1	350	0.217	0.302	0.348	0.303	0.250
LPM1	1750	0.139	0.184	0.215	0.189	0.166
LPM1	3500	0.109	0.151	0.180	0.153	0.143
LPM2	350	0.214	0.298	0.349	0.301	0.257
LPM2	1750	0.134	0.179	0.220	0.191	0.168
LPM2	3500	0.110	0.148	0.182	0.151	0.142
LCM1	350	0.202	0.278	0.318	0.274	0.239
LCM1	1750	0.124	0.169	0.194	0.172	0.149
LCM1	3500	0.105	0.141	0.160	0.136	0.135
LCM2	350	0.207	0.269	0.286	0.258	0.231
LCM2	1750	0.121	0.156	0.173	0.160	0.147
LCM2	3500	0.104	0.134	0.150	0.129	0.125

As shown, spatial sampling methods can be used efficiently in this context of study. Results are encouraging for all sample size and all algorithms used. Indeed, if the sample is spatially balanced, then any not planned domain regularly shaped will have an expected sample size with a very low variance, i.e., it would be approximately fixed [27]. These results confirm the possibility to use spatial sampling methods, whenever the spatial information is available, to produce estimates that are more efficient with respect to use of non-spatial methods, even for domains that are over-represented. The results are remarkable for the estimation of population count in a small geographic area, although strictly related to the population characteristics.

TABLE 10.4

Relative efficiency of the estimation of population size in small geographical domains with unequal inclusion probability (*continued*).

Design	n	Ferrara	Ravenna	Forlì-Cesena	Rimini
SCPS	350	0.260	0.284	0.272	0.233
SCPS	1750	0.175	0.181	0.171	0.138
SCPS	3500	0.148	0.150	0.153	0.121
LPM1	350	0.249	0.273	0.261	0.226
LPM1	1750	0.168	0.180	0.168	0.127
LPM1	3500	0.140	0.150	0.144	0.116
LPM2	350	0.247	0.262	0.253	0.223
LPM2	1750	0.174	0.183	0.176	0.132
LPM2	3500	0.138	0.147	0.139	0.110
LCM1	350	0.245	0.253	0.236	0.205
LCM1	1750	0.159	0.174	0.163	0.119
LCM1	3500	0.132	0.147	0.136	0.106
LCM2	350	0.240	0.250	0.227	0.201
LCM2	1750	0.151	0.156	0.155	0.115
LCM2	3500	0.127	0.136	0.126	0.095

10.5 Business surveys in the presence of locational errors

As explained in the introduction, locational errors could be an important source of bias in sampling studies on business units[3]. Typically, locational errors could be of a dual nature: intentional or unintentional. The first case occurs when data are affected by errors due to privacy protection, so registers suffer by external sources of errors about location of units. In some contexts, such as data about healthcare, forests or individuals, the privacy protection of respondents is a central issue, due to the confidential nature of information and the disclosure risk (i.e., the probability to reveal information about units) that must be avoided [35]. For example, through reverse geo-coding, GIS softwares allow to trace addresses of individuals and then connect sensitive information about them [12, 16]. To reduce the disclosure risk, a geo-masking approach is generally used, in which real geographic positions are randomly misplaced in order to minimize the possibility to identify a unit while, at the same time, preserving the spatial distribution of the variables of interest. The main geo-masking methods consist of swapping, truncating and displacing the geographic coordinates of units [1]. These methods are commonly used in forestry studies, in which a small proportion of sampled

---

[3]Note that in this study we treat units as point-level objects.

lands or trees are randomly misplaced in order to preserve information about values of an acreage or its property.

On the contrary, in the context of establishments data, locational errors are mostly induced by imperfection in the phase of data capture through GPS devices, especially in those zones that are not perfectly covered by satellite positions (e.g., isolated lands or mountains). Usually, when it is not possible to locate the units precisely on the territory, they are positioned on the centroids of the sub-areas that include them [36, 63]. In case of business registers, locations have different levels of precision, which can be classified as: (i) *high*, i.e., coordinates identify the exact address; (ii) *medium*, i.e., coordinates identify the centroid of postal code or census enumeration areas; (iii) *low*, i.e., coordinates identify the centroid of the municipality [15].

Here we concentrate the attention on these unintentional locational errors. In order to investigate the phenomenon of incorrect and incomplete geo-coding of a partition of units and their effects on sample selection, we conduct a study through some simulated scenarios on a spatial population. We consider a simulated population of 1500 units which mimics a dataset coming from Italian business register, for which geographical locations have been randomly and independently generated on a unit square. Each unit is provided by information about geographic position in the space and by a vector of a target variable Y under estimation. The values of the target variable have been generated according to a log-Gaussian random field with mean zero, variance 1, and exponential covariance function

$$\rho(d) = exp\left(\frac{-d}{0.75}\right)$$

where d is the Euclidean distance. The resulting population is represented in Figure 10.2.

A strong spatial cluster in the upper-left section of the figure characterizes the population. Bigger points represent units with higher values of the target variable, and vice versa.

The generated population is the ideal situation, with all locations perfectly assigned to a unique point. A common real situation is that expressed before, in which for some units the exact location is unknown and hence they are assigned to the centroid of their sub-area. Therefore, in order to simulate a realistic situation, we perturbed a proportion of units in three different ways (*Case 1*, *Case 2* and *Case 3*). For each case, the proportion of coarsened locations vary from 5% to 50%, by steps of 0.10. Figures 10.3, 10.4 and 10.5 show, respectively, the three situations, at a 10% level of perturbed locations. The first case (*Case 1*, Figure 10.3) simulates a situation in which units affected by locational errors tend to concentrate in a zone of the study area where they have relatively low values of the target variable and are collapsed on the same location.

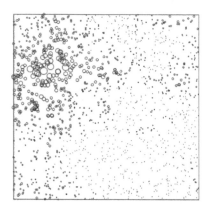

FIGURE 10.2
Random population generated by a log-Gaussian random field.

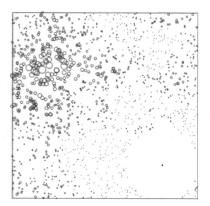

FIGURE 10.3
Case 1: percentage of imprecise locations equal to 0.10 in the bottom-right area of the unit square.

Instead, the second case (*Case 2*, Figure 10.4) is characterized by locational errors in the center of the area under observation, where units can have both high and low values of the target variable.

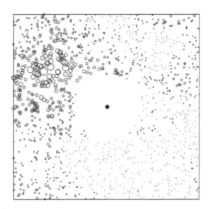

FIGURE 10.4
Case 2: percentage of imprecise locations equal to 0.10 in the central area of the unit square.

Lastly, the third case (*Case 3*, Figure 10.5) simulates the presence of locational errors in the zone where units have relatively high values of the variable under estimation.

For each situation, we have applied, and then compared, the spatial sampling designs presented in Section 10.3, that is the Local Pivotal Methods (LPM1 and LPM2, [28]), the Spatial Correlated Poisson Sampling (SCPS, [27]), the Balanced Sampling by means of Cube method (BSC, [19]) and the Local Cube Method (LCM, [29]). In case of BSC, geographic coordinates of points have been used as vectors of auxiliary information, and in case of LCM, they have been used both as auxiliary variables and as balancing variables. In addition, we implemented the simple random sampling without replacement (SRSWOR) on the starting population and it has been used as a benchmark to evaluate the effect of the use of geographic coordinates in sampling strategies. The above-mentioned sampling designs have been implemented in the context of an equal inclusion probability function. Each design has been used to draw 10000 samples of size equal to 50, 150 and 250 units.

It is assumed that the goal is to estimate the population total of the target variable Y. In order to achieve this aim, we have computed the

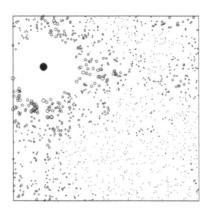

FIGURE 10.5
Case 3: percentage of imprecise locations equal to 0.10 in the upper-left area
of the unit square.

Horvitz-Thompson estimator [34]

$$\hat{Y}_{HT} = \sum_{i=1}^{n} \frac{y_i}{\pi_i}$$

where y_i are the values of the target variable in the population and π_i are the
values of the inclusion probabilities of sample units i. Results of simulation
study are shown in Tables 10.5-10.8.

The comparison between sampling designs have been done in two phases.
Firstly, the relative Root Mean Square Error (rRMSE) of all sampling methods
has been computed for all sample sizes and for all the different scenarios of
"dirty" locations. The rRMSE has been estimated as

$$rRMSE = \frac{\sqrt{\sum_{nsim} \left(\hat{Y}_i - Y_i \right)^2 / nsim}}{Y_i}$$

where Y_i represents the total of the target variable in the population and
nsim represents the number of Monte Carlo repetitions [42], which are, in
the present case, equal to 10000 random samples drawn.

To operate a comparison between the different spatial algorithms and to
evaluate the advantages of including spatial coordinates in the selection procedure of units, the gain in efficiency of all designs has been computed with

respect to SRSWOR. The comparison has been made for all spatial sampling designs, the three different scenarios of positional inaccuracy, the three sample sizes and the ten different proportions of incorrect locations. The gain in efficiency could be expressed as

$$EffGain = \left(1 - \left(\frac{rRMSE(LocError)_{spatialdesign}}{rRMSE(LocError)_{SRSWOR}}\right)\right)100\%$$

which represents the percentage improvement in terms of efficiency, with respect to simple random sampling. Results are shown in Tables 10.5-10.8.

TABLE 10.5

Gain in efficiency (in %) of spatial sampling designs with respect to SRSWOR.

Design	Clean population		
	$n = 50$	$n = 150$	$n = 250$
LPM1	53.07	62.85	65.74
LPM2	54.32	62.41	66.41
SCPS	54.59	63.65	67.11
BSC	23.47	23.24	23.27
LCM	51.11	60.82	64.71

TABLE 10.6

Gain in efficiency (in %) of spatial sampling designs with respect to SRSWOR. Case 1.

Design	Case 1 – 0.10			Case 1 – 0.20			Case 1 – 0.30		
	n=50	n=150	n=250	n=50	n=150	n=250	n=50	n=150	n=250
LPM1	52.82	62.13	65.41	53.10	61.23	64.98	53.47	60.96	64.59
LPM2	52.56	61.95	65.87	52.69	61.87	65.65	52.65	61.46	64.64
SCPS	53.98	63.51	67.38	53.17	63.00	66.06	53.22	61.89	65.23
BSC	24.68	22.60	25.09	23.52	24.93	23.42	25.44	24.33	25.09
LCM	50.85	60.83	64.55	51.70	59.79	64.78	51.56	59.91	63.15

Design	Case 1 – 0.40			Case 1 – 0.50		
	n=50	n=150	n=250	n=50	n=150	n=250
LPM1	51.26	57.94	60.87	45.87	52.52	54.67
LPM2	51.20	57.73	61.43	46.61	52.58	55.20
SCPS	51.14	58.80	61.84	46.54	52.20	54.95
BSC	25.16	25.57	25.63	23.34	22.87	25.05
LCM	49.28	57.27	59.71	44.93	51.24	53.53

As shown, the main general evidence is that when the sample size increases, then the gain in efficiency increases. For all simulations on the clean population, the gain in efficiency is high: it goes from 23% of BSC up to 60% of all other spatial designs. In particular, the best result has been obtained by

TABLE 10.7

Gain in efficiency (in %) of spatial sampling designs with respect to
SRSWOR. Case 2.

	Case 2 – 0.10			Case 2 – 0.20			Case 2 – 0.30		
Design	$n=50$	$n=150$	$n=250$	$n=50$	$n=150$	$n=250$	$n=50$	$n=150$	$n=250$
LPM1	50.75	58.38	62.14	45.64	53.28	55.82	39.22	43.54	46.58
LPM2	51.14	58.96	62.09	46.50	52.35	56.19	38.44	44.58	46.43
SCPS	51.10	59.76	62.91	46.24	52.79	55.74	39.56	44.13	46.29
BSC	21.08	21.29	23.65	19.72	20.21	20.51	17.22	18.12	18.24
LCM	50.75	58.38	62.14	45.64	53.28	55.82	37.74	42.58	45.17

	Case 2 – 0.40			Case 2 – 0.50		
Design	$n=50$	$n=150$	$n=250$	$n=50$	$n=150$	$n=250$
LPM1	29.82	33.51	34.13	19.36	21.96	22.38
LPM2	29.85	32.94	35.21	18.58	21.33	24.42
SCPS	29.01	32.81	34.48	18.58	21.32	22.80
BSC	13.57	13.70	15.49	10.07	10.95	11.42
LCM	28.43	31.69	34.95	17.52	21.52	21.76

TABLE 10.8

Gain in efficiency (in %) of spatial sampling designs with respect to
SRSWOR. Case 3.

	Case 3 – 0.10			Case 3 – 0.20			Case 3 – 0.30		
Design	$n=50$	$n=150$	$n=250$	$n=50$	$n=150$	$n=250$	$n=50$	$n=150$	$n=250$
LPM1	47.85	62.13	54.86	38.95	61.23	43.95	33.90	35.91	36.43
LPM2	48.52	53.00	54.95	39.45	42.25	44.45	33.28	35.78	37.42
SCPS	47.48	51.16	53.43	38.59	41.72	43.71	32.95	35.04	36.69
BSC	23.59	23.66	24.86	24.06	24.13	24.98	23.85	23.79	24.74
LCM	47.85	62.13	54.86	38.95	61.23	43.95	33.94	35.34	36.15

	Case 3 – 0.40			Case 3 – 0.50		
Design	$n=50$	$n=150$	$n=250$	$n=50$	$n=150$	$n=250$
LPM1	35.91	36.43	31.40	36.43	31.40	31.31
LPM2	35.78	37.42	29.85	37.42	29.85	32.89
SCPS	35.04	36.69	29.56	36.69	29.56	31.70
BSC	23.79	24.74	22.40	24.74	22.40	23.04
LCM	35.34	36.15	30.96	36.15	30.96	30.52

SCPS that, with a sample size equal to 250 units, has a gain in efficiency up to
67%. This evidence confirms the consideration about the benefit of the use of
spatial information about locations of units on sample selection procedure,
which lead to estimates that are more efficient.

Turning now to the situations of inexact locations, it can be noted that the
gain in efficiency remains very high for all cases with respect to SRSWOR.
Indeed, higher is the level of inaccurate locations, lower is the gain. It is
important to note that, also in case of 50% of "dirty" locations, the gain in
efficiency remains around a level of 20% in Case 2 and Case 3, and around

50% in Case 1. In all situations, BSC has the lowest performances, due to the balancing procedure of the algorithm [14] and due to the use of spatial coordinates as auxiliary variables (which means that a distance function is not properly computed as in the other spatial algorithms). For the other algorithms, underlined relevant differences between the three mechanisms of arbitrary assignments of coordinates in one point are not detected. However, the gain is lower when the inaccuracy concerns points with a higher value of the target variable, due to the bigger contribution that these points give to the estimation of the total. A particular situation is given by Case 2, in which the coarsened locations are positioned in the middle of the unit square under study. This fact influences in the same way both units with high values of Y and units with low values of Y and it is more similar to a real situation, in which firms with inaccurate locations are located in the centroid of the census area or, in the worst case, in the centroid of the municipality [15].

In conclusion, we can draw two indications. First, as expressed in [21], the use of spatial sampling designs is in any case more advisable with respect to the use of a-spatial designs, to achieve better estimations of the total of a population. Second, it is preferable also in case of a portion of locations assigned in an arbitrary way to a single point, due to spatial uncertainty.

10.6 Conclusions

Recent progresses in geo-referencing procedures have imported knowledge about location of population units in the space, opening the floodgates to several studies that involve positions and behaviors of units on a territory. Together with them, spatial sampling methodologies have experienced flourishing developments in methodologies and computational implementation, making it possible to conduct survey studies on populations provided by geographical information. A field of research in which these progresses are particularly relevant is official statistics context. However, real data coming from official archives are often affected by errors, which can concern, among others, the completeness of registers and the locational accuracy of units. In this chapter, we have focused the attention on these issues, by analyzing two cases, in which it is possible to use spatial sampling methods without neglecting errors of archives.

In the first study, we face the problem of incompleteness of registers and, in particular, with coverage errors. We analyze the case of Italian Population and Housing Census and its problems, deriving from costs and time features, which can lead to serious under-coverage errors. Then, we have introduced capture-recapture models, which are used in some cases to estimate the unknown size of a population, by using information collected by multiple lists. By means of a simulation study on provinces of Emilia-Romagna region in

Italy, the aim of this study was twofold. First, we have estimated coverage errors through spatial sampling methods, assessing the important role of these methodologies in the context of study, both for equal and unequal inclusion probabilities, and we have compared them with a-spatial designs. Second, we have showed the important role of spatial sampling designs in the problem of estimating coverage rate for not planned small geographic areas within a region, by using the case of nine provinces in Emilia Romagna.

In the second study, we have considered the case about locational errors of units. In some archives, such as for example business or health registers, errors about position of units on the territory are ordinary, due to privacy issues and to errors in geo-coding procedures. This second case is relevant in business registers, because of imperfection in data capture in difficult areas and of movement in market, which can lead to a quick change of position of establishments. We have proceeded to conduct a simulation study on a generated population that mimics a business population coming from official registers. Then, we have implemented spatial sampling by means of recent algorithms, in order to assess the efficiency of these methodologies, both in case of correct and with different percentage of incorrect locations of units. We have compared spatial methods with simple random sampling without replacement, in case of equal inclusion probabilities, for total population estimation.

For both applications, results are univocal in considering spatial sampling methodologies to be more efficient with respect to a-spatial methods, leading the conclusion that the use of geographical information in sampling, when available, is more advisable for official statistics studies.

Bibliography

[1] William B Allshouse, Molly K Fitch, Kristen H Hampton, Dionne C Gesink, Irene A Doherty, Peter A Leone, Marc L Serre, and William C Miller. Geomasking sensitive health data and privacy protection: an evaluation using an e911 database. *Geocarto international*, 25(6):443–452, 2010.

[2] Luc Anselin. Thirty years of spatial econometrics. *Papers in regional science*, 89(1):3–25, 2010.

[3] Luc Anselin. *Spatial econometrics: methods and models*, volume 4. Springer Science & Business Media, 2013.

[4] Giuseppe Arbia. The use of gis in spatial statistical surveys. *International Statistical Review/Revue Internationale de Statistique*, pages 339–359, 1993.

[5] Giuseppe Arbia and Giuseppe Espa. *Statistica economica territoriale.* Cedam, 1996.

[6] Roberto Benedetti, Giuseppe Espa, and Emanuele Taufer. Model-based variance estimation in non-measurable spatial designs. *Journal of Statistical Planning and Inference*, 181:52–61, 2017.

[7] Roberto Benedetti and Federica Piersimoni. A spatially balanced design with probability function proportional to the within sample distance. *Biometrical Journal*, 59:1067–1084, 2017.

[8] Roberto Benedetti, Federica Piersimoni, and Paolo Postiglione. *Sampling Spatial Units for Agricultural Surveys.* Springer, 2015.

[9] Roberto Benedetti, Federica Piersimoni, and Paolo Postiglione. Spatially balanced sampling: a review and a reappraisal. *International Statistical Review*, 85(3):439–454, 2017.

[10] Lennart Bondesson and Daniel Thorburn. A list sequential sampling method suitable for real-time sampling. *Scandinavian Journal of Statistics*, 35(3):466–483, 2008.

[11] Maged N Kamel Boulos. Towards evidence-based, gis-driven national spatial health information infrastructure and surveillance services in the United Kingdom. *International Journal of Health Geographics*, 3(1):1, 2004.

[12] John S Brownstein, Christopher A Cassa, and Kenneth D Mandl. No place to hide reverse identification of patients from published maps. *New England Journal of Medicine*, 355(16):1741–1742, 2006.

[13] Burrough, PA, McDonnell, R, McDonnell, RA, and Lloyd, CD. *Principles of Geographical Information Systems.* Oxford university press, 2015.

[14] Guillaume Chauvet and Yves Tillé. A fast algorithm for balanced sampling. *Computational Statistics*, 21(1):53–62, 2006.

[15] M Cozzi and D Filipponi. The new geospatial business register of local units: potentiality and application areas. In *3rd Meeting of the Wiesbaden Group on Business Registers-International Roundtable on Business Survey Frames, Washington, DC*, pages 17–20, 2012.

[16] Andrew J Curtis, Jacqueline W Mills, and Michael Leitner. Spatial confidentiality and gis: re-engineering mortality locations from published maps about hurricane katrina. *International Journal of Health Geographics*, 5(1):44, 2006.

[17] JC Deville, JM Grosbras, and N Roth. Efficient sampling algorithms and balanced samples. In *Compstat*, pages 255–266. Springer, 1988.

[18] Jean-Claude Deville and Yves Tillé. Unequal probability sampling without replacement through a splitting method. *Biometrika*, 85(1):89–101, 1998.

[19] Jean-Claude Deville and Yves Tillé. Efficient balanced sampling: the cube method. *Biometrika*, 91(4):893–912, 2004.

[20] Loredana Di Consiglio and Tiziana Tuoto. Coverage evaluation on probabilistically linked data. *Journal of Official Statistics*, 31(3):415–429, 2015.

[21] Maria Michela Dickson, Roberto Benedetti, Diego Giuliani, and Giuseppe Espa. The use of spatial sampling designs in business surveys. *Open Journal of Statistics*, 4(05):345, 2014.

[22] Maria Michela Dickson and Yves Tillé. Ordered spatial sampling by means of the traveling salesman problem. *Computational Statistics*, 31(4):1359–1372, 2016.

[23] PJ Diggle. Statistical analysis of spatial point patterns, London: Arnold. *MATH Google Scholar*, 2003.

[24] Stephen E Fienberg. Bibliography on capture-recapture modelling with application to census undercount adjustment. *Survey Methodology*, 18(1):143–154, 1992.

[25] Carlo Gaetan and Xavier Guyon. *Spatial statistics and modeling*, volume 81. Springer, 2010.

[26] A Grafström and J Lisic. Balanced sampling: Balanced and spatially balanced sampling. *R package version*, 1(2), 2016.

[27] Anton Grafström. Spatially correlated poisson sampling. *Journal of Statistical Planning and Inference*, 142(1):139–147, 2012.

[28] Anton Grafström, Niklas LP Lundström, and Lina Schelin. Spatially balanced sampling through the pivotal method. *Biometrics*, 68(2):514–520, 2012.

[29] Anton Grafström and Yves Tillé. Doubly balanced spatial sampling with spreading and restitution of auxiliary totals. *Environmetrics*, 24(2):120–131, 2013.

[30] Moshe Haspel and H Gibbs Knotts. Location, location, location: Precinct placement and the costs of voting. *The Journal of Politics*, 67(2):560–573, 2005.

[31] AS Hedayat and Dibyen Majumdar. Generating desirable sampling plans by the technique of trade-off in experimental design. *Journal of Statistical Planning and Inference*, 44(2):237–247, 1995.

[32] AS Hedayat, CR Rao, and J Stufken. 24 designs in survey sampling avoiding contiguous units. *Handbook of statistics*, 6:575–583, 1988.

[33] AS Hedayat, CR Rao, and J Stufken. Sampling plans excluding contiguous units. *Journal of Statistical Planning and Inference*, 19(2):159–170, 1988.

[34] Daniel G Horvitz and Donovan J Thompson. A generalization of sampling without replacement from a finite universe. *Journal of the American statistical Association*, 47(260):663–685, 1952.

[35] Anco Hundepool, Josep Domingo-Ferrer, Luisa Franconi, Sarah Giessing, Rainer Lenz, Jane Longhurst, E Schulte Nordholt, Giovanni Seri, and P Wolf. Handbook on statistical disclosure control. *ESSnet on Statistical Disclosure Control*, 2010.

[36] Geoffrey M Jacquez. A research agenda: does geocoding positional error matter in health gis studies? *Spatial and spatio-temporal epidemiology*, 3(1):7–16, 2012.

[37] Matthew Jaro. Record linkage research and the calibration of record linkage algorithms. In *Statistical Research Division Report Series SRD Report No. Census/SRD/RR-84/27*. Citeseer, 1984.

[38] Andre G Journel and Ch J Huijbregts. *Mining geostatistics*. Academic Press, 1978.

[39] BN Mandal, Rajender Parsad, VK Gupta, and UC Sud. A family of distance balanced sampling plans. *Journal of Statistical Planning and Inference*, 139(3):860–874, 2009.

[40] Kenneth H Pollock. Review papers: modeling capture, recapture, and removal statistics for estimation of demographic parameters for fish and wildlife populations: past, present, and future. *Journal of the American Statistical Association*, 86(413):225–238, 1991.

[41] Jerry H Ratcliffe. Geocoding crime and a first estimate of a minimum acceptable hit rate. *International Journal of Geographical Information Science*, 18(1):61–72, 2004.

[42] Christian Robert and George Casella. Monte Carlo statistical methods. Springer-Verlag. *New York*, 2004.

[43] Richard M Royall and Jay Herson. Robust estimation in finite populations i. *Journal of the American Statistical Association*, 68(344):880–889, 1973.

[44] Gerard Rushton, Marc P Armstrong, Josephine Gittler, Barry R Greene, Claire E Pavlik, Michele M West, and Dale L Zimmerman. Geocoding in cancer research: a review. *American Journal of Preventive Medicine*, 30(2):S16–S24, 2006.

[45] Carl J Schwarz and George AF Seber. Estimating animal abundance: review iii. *Statistical Science*, pages 427–456, 1999.

[46] George AF Seber. A review of estimating animal abundance. *Biometrics*, pages 267–292, 1986.

[47] George AF Seber. A review of estimating animal abundance ii. *International Statistical Review/Revue Internationale de Statistique*, pages 129–166, 1992.

[48] Don L Stevens and Anthony R Olsen. Variance estimation for spatially balanced samples of environmental resources. *Environmetrics*, 14(6):593–610, 2003.

[49] Don L Stevens Jr and Anthony R Olsen. Spatially balanced sampling of natural resources. *Journal of the American Statistical Association*, 99(465):262–278, 2004.

[50] A Stewart Fotheringham and Peter A Rogerson. Gis and spatial analytical problems. *International Journal of Geographical Information Science*, 7(1):3–19, 1993.

[51] J Stufken. *Combinatorial and statistical aspects of sampling plans to avoid the selection of adjacent units*. Iowa State University. Department of Statistics. Statistical Laboratory, 1991.

[52] John Stufken, Sung Y Song, Kyoungah See, and Kenneth R Driessel. Polygonal designs: Some existence and non-existence results. *Journal of statistical planning and inference*, 77(1):155–166, 1999.

[53] Y Tillé and A Matei. The R package sampling. The comprehensive R archive network, 2005.

[54] Graham Upton, Bernard Fingleton, et al. *Spatial data analysis by example. Volume 1: Point pattern and quantitative data*. John Wiley & Sons Ltd., 1985.

[55] William E Winkler. The state of record linkage and current research problems. In *Statistical Research Division, US Census Bureau*. Citeseer, 1999.

[56] Kirk M Wolter. Some coverage error models for census data. *Journal of the American Statistical Association*, 81(394):337–346, 1986.

[57] James H Wright and John Stufken. New balanced sampling plans excluding adjacent units. *Journal of Statistical Planning and Inference*, 138(11):3326–3335, 2008.

[58] Duck-Hye Yang, Lucy Mackey Bilaver, Oscar Hayes, and Robert Goerge. Improving geocoding practices: evaluation of geocoding tools. *Journal of medical systems*, 28(4):361–370, 2004.

[59] Frank Yates. A review of recent statistical developments in sampling and sampling surveys. *Journal of the Royal Statistical Society*, 109(1):12–43, 1946.

[60] PSF Yip, G Bruno, N Tajima, GAF Seber, ST Buckland, RM Cormack, N Unwin, YF Chang, SE Fienberg, BW Junker, et al. Capture-recapture and multiple-record systems estimation: history and theoretical development. *American Journal of Epidemiology*, 1995.

[61] Paul A Zandbergen. A comparison of address point, parcel and street geocoding techniques. *Computers, Environment and Urban Systems*, 32(3):214–232, 2008.

[62] Li-Chun Zhang. On modelling register coverage errors. *Journal of Official Statistics*, 31(3):381–396, 2015.

[63] Dale L Zimmerman. Estimating the intensity of a spatial point process from locations coarsened by incomplete geocoding. *Biometrics*, 64(1):262–270, 2008.

Index

Note: Page numbers followed by "*n*" with numbers indicate footnotes.